ADVICE TO THE ADVENTUROUS

As a young lad, attempting to earn a living doing man's work, fear of failure was constant. The fear generated desperation plus sufficient stick-to-itiveness to produce a small measure of success. Through the years, I learned that fear is one of life's most important stimulants, a power than can help conceive and produce some of life's richest dreams.

Building a house may seem like a big deal, a frightening voyage into the unknown, but in reality it only requires digging one shovel full of earth, sawing one 2 x 4, or driving one nail at a time. It does require direction, the kind this book provides.

The author assumes you have never built a house and explains HOW — in non-technical language everyone can follow. Read this book through several times. Ask your lumber dealer to explain any step you can't figure out. Then ask yourself — if others can, why can't you?

Every year over 11% of all houses constructed are built by people in their spare time for their own occupancy. Colonists who built this nation built their own homes. Doing something you have never done before allows you to accurately measure a person called YOU.

Don R. Brann

TABLE OF CONTENTS

14 – Facts About Buying Land
25 – Site Selection
30 – List of Material
34 – Lay Out Guide Lines
40 – Footings
44 – Excavate for a Full Basement
49 – Foundation Walls
58 – Termite Shield
63 – Porch Foundation
66 – Basement Partitions
77 – Floor Joists
80 – Subflooring
81 – First Floor Wall Frames
98 – Ridge and Rafters
108 – Frame for Louver
109 – Plywood Sheathing
114 – Roofing
120 – Gutters and Leaders
122 – Install Windows
123 – Install Chimney
125 – Rough in Plumbing, Wiring, Heating
130 – Exterior Siding and Porch Ceiling
132 – Hang Exterior Doors
138 – Basement Floor
140 – Facts About Stair Construction
150 – Insulation
153 – Kitchen Planning
154 – Finished Flooring
155 – Hang Doors, Trim
158 – Optional Construction
168 – Truss Rafters
169 – Steel Framing Construction
171 – Two Story House
174 – Facts About Wiring
178 – Facts About a Septic Tank
187 – The Foundation House
196 – Opportunity Unlimited
202 – Cross Reference Index
216 – Other Books By Author

how to build a LOW COST HOUSE ABOVE or BELOW GROUND

DONALD R. BRANN

FIFTH PRINTING — 1982
REVISED EDITION

Published by
EASI-BILD DIRECTIONS SIMPLIFIED, INC.
Briarcliff Manor, NY 10510

Library of Congress Card #81-69875

FIRST PRINTING

REVISED EDITIONS
1970,1972,1975,1982

ISBN 0-87733-832-9

NOTE:
All metric dimensions are shown within 5/100 of a centimeter.

Due to the variance in quality and availability of many materials and products, always follow directions a manufacturer and/or retailer offers. Unless products are used exactly as the manufacturer specifies, its warranty can be voided. While the author mentions certain products by trade name, no endorsement or end use guarantee is implied. In every case the author suggests end uses as specified by the manufacturer prior to publication.

Since manufacturers frequently change ingredients or formula and/or introduce new and improved products, or fail to distribute in certain areas, trade names are mentioned to help the reader zero in on products of comparable quality and end use.

The Publisher

READ, LEARN, SAVE

Years ago an elderly neighbor offered a bit of advice I have never forgotten. "Never buy a house until you learn how to appraise its value; never buy land until you find out what you are getting; never buy a business until you have worked for the owner long enough to learn how to operate it profitably."

Through the years, I've known people in every age and income group who made mistakes they couldn't afford because they failed to follow this advice. They bought houses because they "needed a place to live," then found themselves saddled with repair costs that forced them to sell at a loss. Others bought land loaded with liens.

Today a new and vitally important element must be seriously considered prior to the purchase of land. Even in seemingly safe neighborhoods, juvenile delinquents, unrestrained by parents, continually wreak havoc. When arrested and released with a reprimand by judges, each makes life even more unbearable for many homeowners.

Others buy a building site not realizing a zoning change will place a highway within feet of their dream home. Before signing any agreement, spend time talking to "locals." Subscribe to a local newspaper or spend a morning going through back issues to better appraise what new zoning changes have been discussed or approved. Remember — the seller is in no way obligated to reveal these facts unless you request this information. He doesn't need to mention he drilled one or two wells and didn't get water.

Even where conditions look acceptable, try stretching out the closing date as long as possible with the smallest down payment you can afford to lose. Where you have time, rent a house in the area for a year prior to building. Why be a loser when only an investment of time can show you where and when to build.

Before signing a house or land sale agreement, read this book. Not once, but two or three times. Make a check list and follow it when shopping for land. Don't become enamored with a view, a salesman or a price. It's far better to lose something you don't own than to get stuck with a lemon.

While people build for many different reasons, relatively few intelligently prepare for a purchase that frequently influences the best years of their lives. Even fewer obtain expert advice. This book can save you thousands of dollars. It can also help transform spare time into thousands of dollars in Capital Gains. Even those who do none of the building quickly learn how to appraise construction, how not to get "conned."

People who build a home for their own occupancy are twice blessed. By transforming spare time into building labor, they create a nest egg, an investment that can produce sizeable Capital Gains. They obtain a home at the lowest possible cost. Equally important, they no longer fear costly repairs or improvements. Having learned to build, repairs are easy.

The growing economic need to save energy and find a "house you can afford," has created a new approach to shelter. Simplified directions explain how to build a one story house over a full basement, Illus. 1. The first floor provides comfortable living space for a single or couple.

The full basement, Illus. 2, with daylight and cross ventilation supplied by window wells, contains space that some owners use as a second and third bedroom.

The first floor could also be built over a crawl space, Illus. 3.

A hilly building site offers a natural setting for both a side hill house, Illus. 4, or an earth sheltered dwelling, Illus. 5.

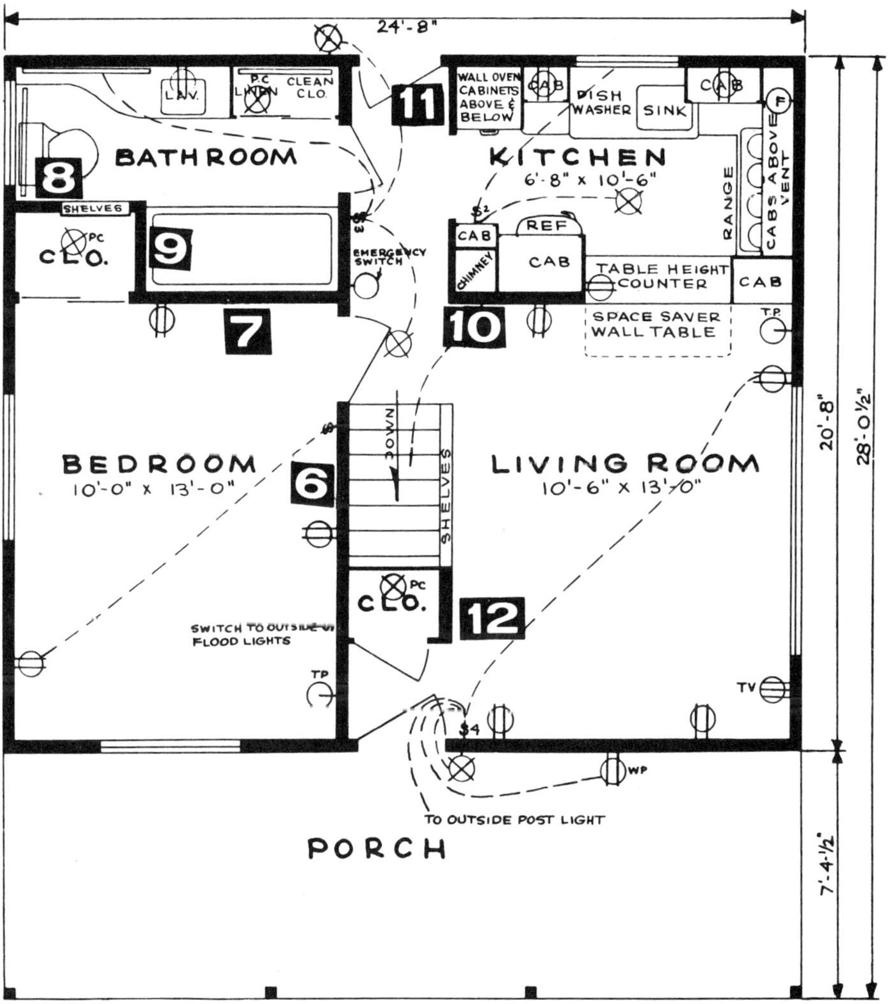

ENTRANCE FLOOR

(1)

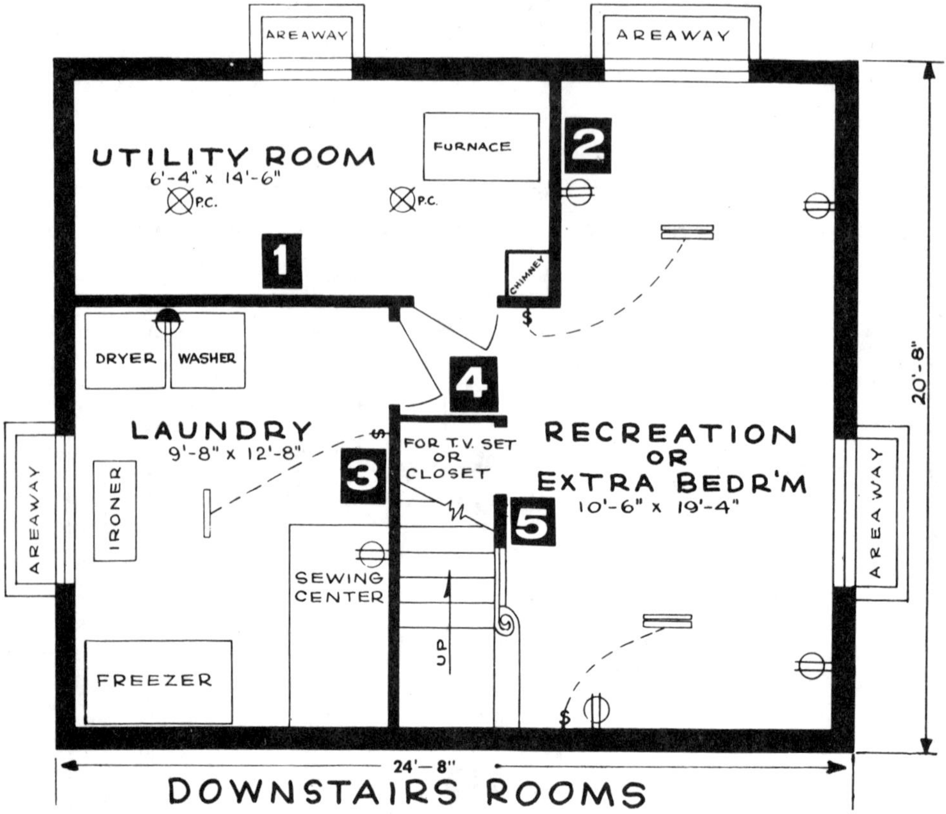
AREAWAY
AREAWAY
UTILITY ROOM
6'-4" x 14'-6"
P.C.
P.C.
FURNACE
2
1
CHIMNEY
DRYER
WASHER
4
LAUNDRY
9'-8" x 12'-8"
FOR T.V. SET
OR
CLOSET
3
RECREATION
OR
EXTRA BEDR'M
10'-6" x 19'-4"
5
AREAWAY
IRONER
SEWING
CENTER
UP
AREAWAY
FREEZER
20'-8"
24'—8"
DOWNSTAIRS ROOMS

2

3

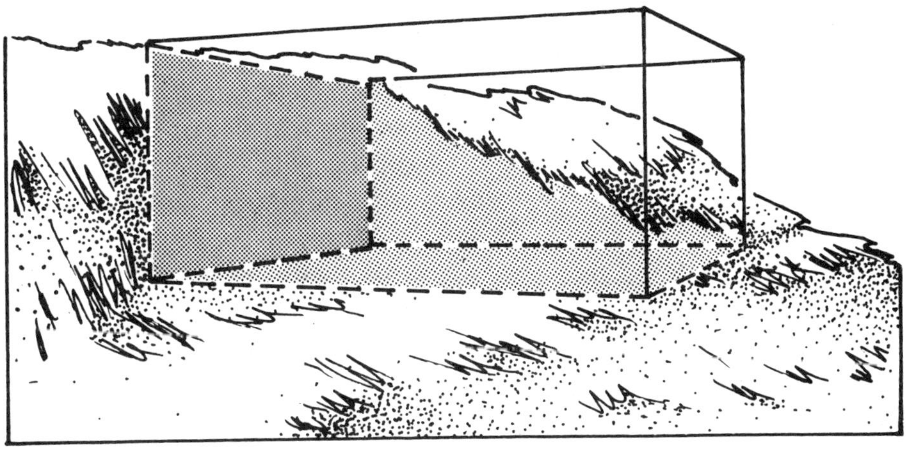

4

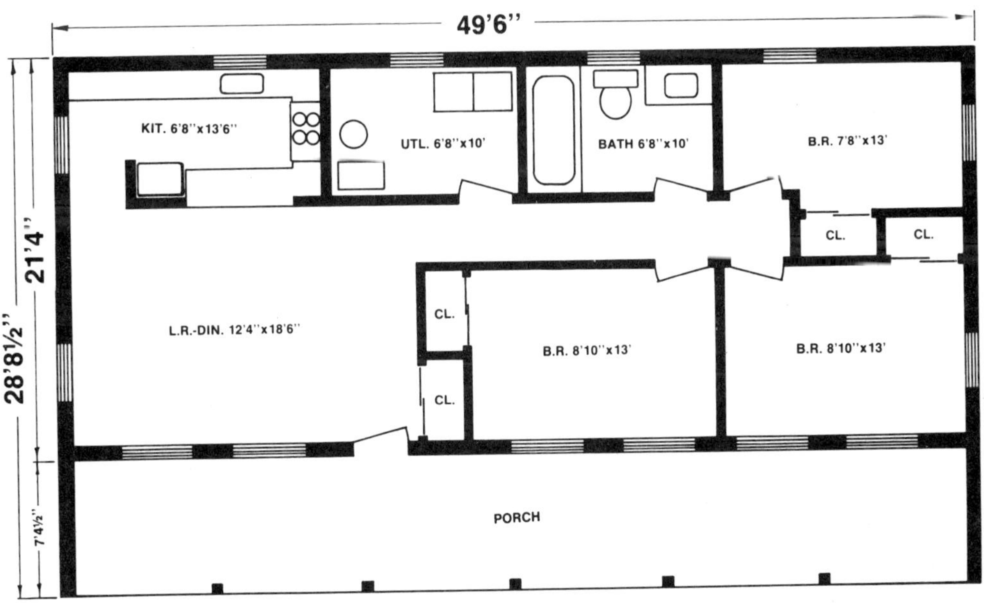

5

EARTH SHELTERED FOUNDATION HOUSE

Use wire reinforced 12" blocks on outside walls, 4" concrete blocks for partitions

A house protected on three sides by earth provides great energy savings. Plan to excavate to a depth so four courses of block are above grade. When the exposed front, Illus. 5, faces south, it simplifies entry. Where winter sunlight can be captured, enclose the porch from floor to ceiling with clear plastic and you generate an enormous supply of free heat.

If the site is level and you want an earth covered roof, entry can be made through a greenhouse built according to directions in Book #811 How to Build a Greenhouse. Stairs at end of greenhouse provide entry. In this construction, use reinforced concrete joists, Illus.155, precast concrete slabs and a wire reinforced concrete deck. This must be waterproofed and insulated prior to covering with earth.

Illus. 5 shows a foundation house. This can be constructed with 12'' concrete block or a poured foundation. Its overall size is equivalent to space shown in Illus. 1 and 2. Following step by step directions you can build any other size.

READ, LEARN, BUILD and SAVE.

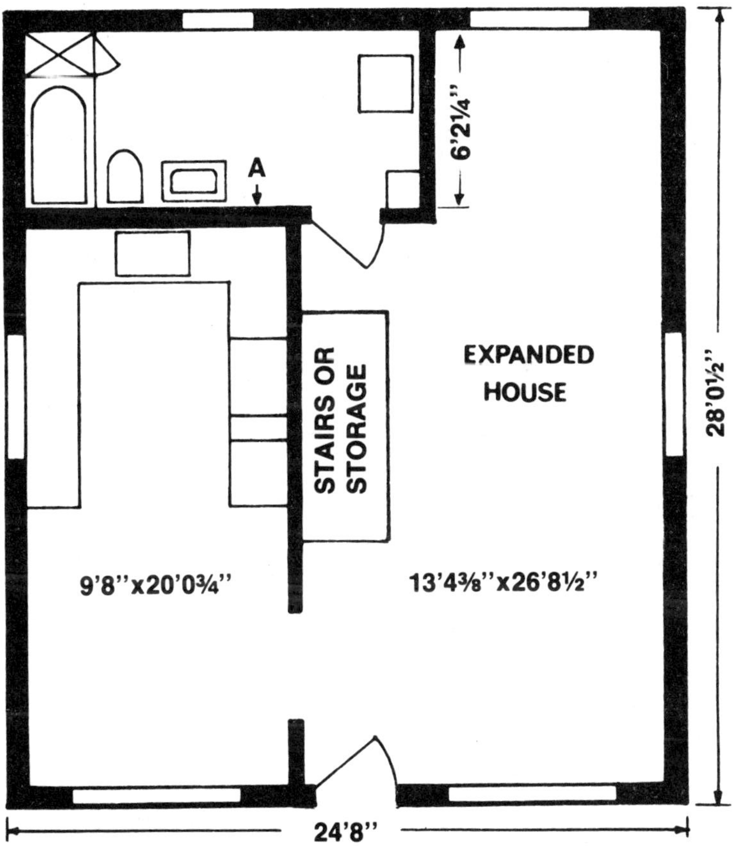

6

Those who want to build a slightly larger house can include the porch area, Illus. 6.

FACTS ABOUT BUYING LAND

Any site selected for building must have an adequate supply of fresh water. Property not within a water district must rely on a well. Talk to neighboring property owners to ascertain depth of their wells, and whether they obtained a sufficient supply during a dry spell. Talk to at least two well diggers to find out what they charge to go through earth and/or rock. Add at least 50% to his estimate. Or better still double the price he quotes. When added to the cost of the site, does it still make a sound investment?

When analyzing whether to buy in an area where there is any question of an adequate water source, note whether neighboring property owners use a washing machine. While hundreds of homes were built in areas without drilling wells, each relied on large cisterns and sufficient annual rainfall to keep them filled. Any change in the weather can create a dust bowl. This requires transporting water from whatever source is available.

Drilling for water is a chancy venture. While past performances in an area provide some indication of a water table, the amount of rock and water at any one level, and the integrity of the driller can create a heartbreaking experience for those operating with limited financial means.

Another important point is your being able to do business with those who have the equipment and experience. The rig required to drill a well costs money to move. Unless you "do business' with available drillers, you can't afford to have a well dug.

Another important point — be sure the driller specifies equipment he will furnish, size of pipe, pump, storage and/or pressure tank, size and kind of pipe he will run into house; also whether he will make the final connection to your house. Many only quote on drilling a well. You have to provide equipment needed to get the water out.

Be sure to ascertain quality of equipment being supplied, what other equipment will be needed, also whether the storage tank specified is sufficient to handle your family needs. Visit a store that sells pumps. They can tell you what size tank you should install. This could be twice the size a "friendly driller" may say you need.

Another important and costly problem concerns burying a water line below frost level. If the site is rocky, this requires costly blasting. On the other hand, you might find it far less costly to lay double thickness rock wool insulation around water line, then cover line with three, or as many feet of earth fill local frost level requires.

When considering a low or rocky site or even the most attractive building site, adequate space for a septic tank and field is a must when there is no town sewer line.

Before you sign any purchase agreement, learn where septic tank permits are issued. Take time to discuss a site for tank and field with the department of health or building inspector. Get their opinion as to exact location a septic tank and field will be allowed or disallowed. If you feel compelled to buy before getting this opinion, make your deposit and/or purchase contingent on getting both a building permit and a permit for a septic tank. Without both, you can't build in many areas supervised by building inspectors.

Property located adjacent to a reservoir may also restrict installation of a septic tank and field.

The cost of excavation must also be considered before a purchase. Note whether any rock project above ground in the area you envision for a building site. If rock projects in a low area, covering with at least three to four feet of sand fill frequently provides an ideal field for a septic tank. The local building or department of health inspector will specify how much fill they require before issuing a permit.

If owner of land permits, drive a crowbar into ground at various points within area selected for house, also area selected for a septic tank. This will give you some idea as to depth of rock below surface.

A building site outside of a water and sewer district must provide sufficient space from house and property line for both the septic tank and drainage field, note page 177. You also need space to bury a fuel tank if local ordinances prohibit it being placed above ground. Since you can't drive over a septic tank or its field, you must consider space for a driveway and garage if same is desired.

If the site is level and a full basement is to be excavated, hire a bulldozer or dig it by hand. The space is too valuable not to use. A full basement or even a half basement is better than none at all, and its square foot cost is usually less than any other.

Some land is frequently made usable by building concrete block or poured concrete foundation walls to the height required, then raising the level of grade against walls to provide a natural runoff.

Don't expect the present owner of a low lying site to tell you how long water remains after a rain. Note slope of surrounding terrain. Will the site you select collect water from the surrounding area?

Land around a lake can flood during a heavy rain and remain dry under normal conditions. If neighboring houses were built on high foundations without basements or on steel columns, the chance of previous flooding required this construction.

Marshy land is usable if there's hardpan below. When in doubt, get permission to dig test holes. Buy a post hole digger and don't hesitate to dig as deep as you plan on laying footings. If you hit a layer of clay, gravel or sand, even if the hole fills with water, oversize footings will invariably support

Use steel girders or two 2x8 or 2x10 depending on span and local code requirements.

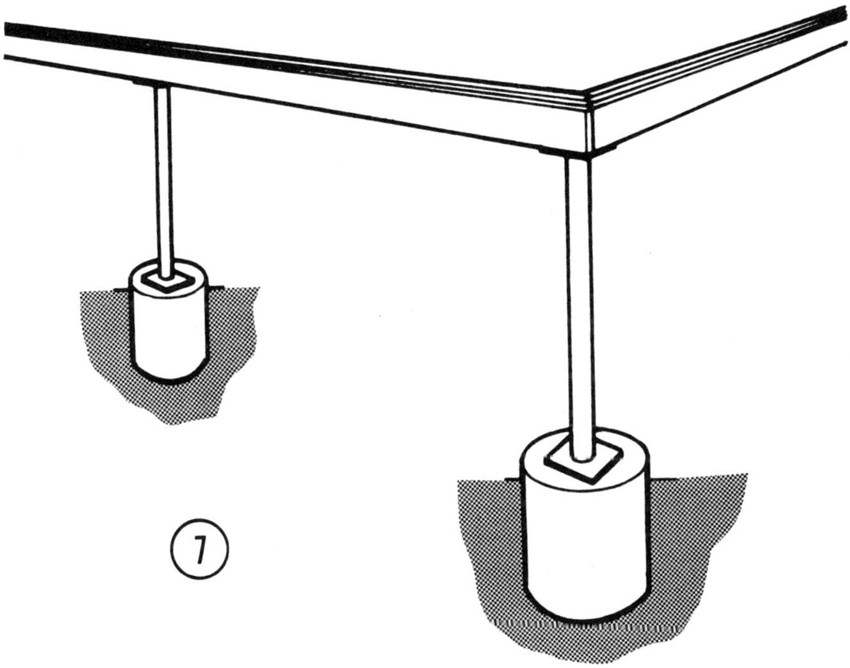

7

a house on steel columns, Illus. 7. In this case, it's worth consulting a local engineer to qualify the site prior to purchase.

If you fall in love with a hilly or rocky site, and the price including estimated cost of a well is right, there is much that can be done without blasting. Footings can be secured to rock. On a hilly site, build walls to slope required, Illus. 8. This can provide a two story house on low side, one story on high side. The added cost of block is small compared to the extra amount of living space.

To secure footings, rent an electric hammer and drill ½" holes 8 to 10" deep, every three to four feet. Cut ½×2'0" reinforcing rod and embed these in holes.

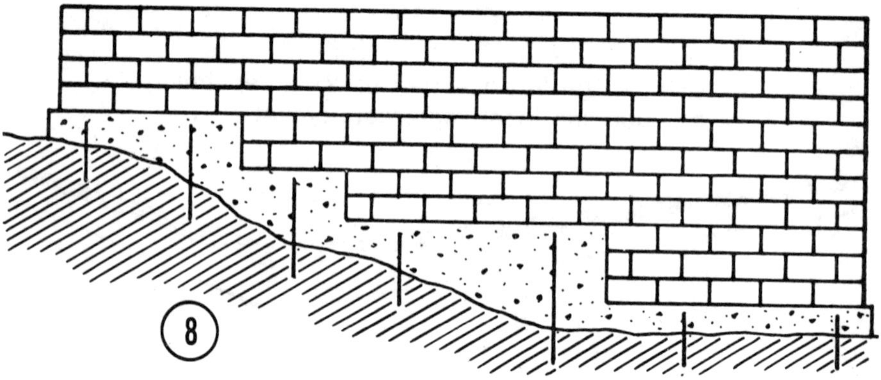

Or you can build footing forms and/or footing and foundation forms and pour readymix concrete, Illus. 9. Use 5/8 or 3/4'' plywood with 2 x 4'' stiffeners. Build forms 16'' wide for 8×8×16''block; 20'' wide if 12×8×16'' block is used.

Many earth sheltered houses are built with 10, 12, 14 or 16'' wide poured concrete foundations reinforced with 1/2'' steel rod.

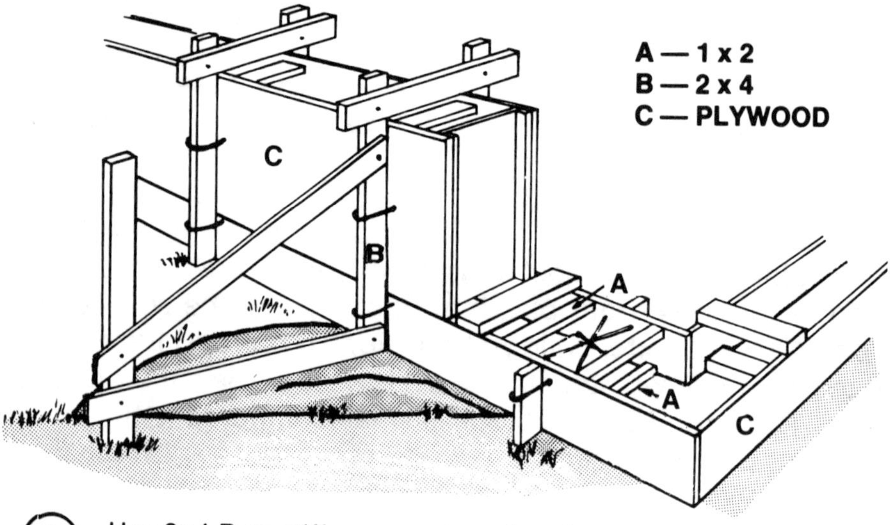

(9) Use 2x4 B as stiffeners on top of rock, or as stakes where ground permits.

REVIEW THESE FACTS

1. Don't sign your name or even initial a sales contract or put a dollar down as a deposit or binder until your lawyer approves the sales contract.

2. Your lawyer can advise whether you can obtain clear title with no right of way or other encumbrances.

3. Does the neighborhood have a future?

4. Is it protected by zoning that prohibits erection of gas stations, factories, etc?

5. Are there any unpaid taxes, mechanics or other liens against the property? What are present tax assessments? What new assessments have already been adopted that will affect the property? The only place to get a complete answer is at the tax assessor's office. They will also tell you how many times, in its recent past, the property has changed hands.

6. Is the property convenient to what you need — schools, churches, shopping center, transit facilities, and place of employment? Is there adequate police and fire protection to permit an acceptable insurance rate?

7. Will you be permitted to build the size house your family requires? Building codes invariably specify how far a house can be constructed from the street or property line. Many people have bought land only to learn they couldn't build the house they wanted because there wasn't sufficient space for a septic tank.

8. Does the land complement the kind of house you want to build?

9. Does it offer a site with natural drainage away, and not toward the house?

10. Does the site provide sufficient space for a house, septic tank and field, garage and outdoor living? Does this space offer the privacy you want? Consider placement of a garage and driveway, its grade and approach to street. Will snow drift into area? Will automobile headlights flash into living or bedroom windows?

Hire a local lawyer to represent you, one not related or indebted to the seller. If he is familiar with the property and its history, he may be able to save you time, trouble and expense. While most lawyers can't give you an exact cost of a closing, they can tell you the many problems that could arise, and what you can expect in the way of costs. Make arrangements to pay your lawyer one fee if you go through with the closing, another fee if the closing reveals conditions that prevent your taking title. It's always better to back out of a deal with a small loss than to go ahead and take a larger one.

Be sure to have a title search made. This is a sound investment. Title insurance protects your investment.

Buying a house or land is frequently a once in a lifetime experience. As such it naturally creates fear and packs an emotional wallop. Even a level headed businessman loses some of his cool when he discovers the amazing number of problems that can arise during a closing. For these reasons, it's important to play safe. Don't sign or agree to anything until you know what you are buying.

A case in point is the size of parcel. Let's say it's offered as an acre. Is the forty odd thousand square feet of land shaped like you think it is, with a wide road frontage, or are you getting fifty foot frontage and an odd shaped parcel that won't complement your proposed building?

Before purchase find out if a lending institution will lend money on the land, and how much of a mortgage they think is sound. Their appraisal will give you some inkling as to what others judge its value. Always talk to the most reputable lending institution. Shady loan companies will lend as much money as they can when they investigate your ability to pay. To them the value of the land is incidental.

If you don't ask questions, the seller is not obligated to bring them up. Unless you ask you could one day be surprised to see a bulldozer carve a street through the middle of your property, or a trenching machine digging a ditch for a sewer line where you wanted to locate your living room. Don't take the owner's word. Get everything in writing. If he's honest, he won't hesitate to write down everything he says.

As every lawyer will advise, make your purchase contingent on:

1. Getting a clear title.

2. That no liens are filed against the property.

3. That purchase is dependent on obtaining a mortgage from an institution of your choice.

4. That the property is shaped the way it is represented.

5. That no right of way or other encumbrances exist.

6. That the seller will vacate the property and will not remove anything you agree must remain.

7. Your being able to obtain a building, septic tank and occupancy permit.

TOOLS

The following tools will be needed. Pick, shovel, hoe, cement mixing tub (making one 3 x 6 x 1' deep), mason's trowel, mason's float, ball of building line, plumb bob, line level, carpenter's level, framing square, miter square, hammer, brace, 9/16, 5/8, 3/4'' bits. Crosscut, Rip and Compass saws. Expansion bit, jack, block and smooth plane, fifty foot steel tape. Chalk line, hatchet, folding zig-zag rule, nailset. 1/4, 1/2, 3/4, 1'' wood chisels, miter box, wrecking bar, butt gauge. Sandpaper or sanding machine, adjustable end wrench, screw driver, two extension clamps. Nailing apron, overalls. 26'' and 5' horses; buy adjustable sawhorse clamps and cut legs to length required.

One straight edge (use a straight piece of 5/4 x 6'' x 8' or a 2 x 4'' x 8'). Step ladder, plus a 12 or 14 ft. ladder. This can be made by nailing 1 x 4 rungs on two 2 x 4's. Space rungs about one foot apart. Nail securely to uprights with 8 penny common nails.

If in doubt about the proper use of any tool, your retailer will explain its function.

In many areas, lumber and building material dealers rent small concrete mixing machines by the day. These machines are great time and labor savers and well worth the cost of hire.

An electric hand or table saw is also an excellent investment. These frequently equal an extra man's time. If you plan on doing any building, the cost of this equipment can be saved many times over during construction. Most electric companies will supply temporary service if you provide a weatherproof housing for the meter. Use underground cable. Mount this to a temporary 4 x 4 post. Leave enough slack to permit mounting meter box on outside of completed house.

Before starting construction, read this book through carefully. Note location of each framing member when mentioned, study each illustration to familiarize yourself with procedure. Always consider yourself outside, facing the completed house, when reference is made to left or right. Also note ceiling height specified in basement and first floor. If you want more ceiling height in basement, add an extra course of block to foundation walls and extra step to stairs.

If you want a higher ceiling on first floor, cut studs to length required.

If you have never done any building, the illustrations will at first seem strange. After going over them two or three times, you will begin to visualize construction. If any step is not completely clear, talk to your lumber dealer. He is familiar with construction and will answer most questions. Once construction starts, each progressive step becomes increasingly clear as a piece of lumber becomes a stud, plate, etc.

While many folks will build this house themselves, others will hire a mason or carpenter to help get started. When you work alongside an experienced workman, you get the job done faster and gain valuable experience.

Should you employ masonry or carpentry help, consult your insurance agent concerning liability and other insurance to properly and adequately protect you, your property and the individual you employ.

CAUTION

Building is fun. It provides healthy relaxation, exercise, a great feeling of accomplishment while it opens your eyes to new fields of endeavor. But remember one thing — work only when you feel like it, and only for as long as you feel physically fit. Once you become tired, accidents and errors occur, particularly when doing something you are normally unaccustomed to doing.

Always get adequate assistance when raising wall frames, etc. Don't overtax your strength.

Actual construction follows this general procedure:

1. Obtain a building and septic tank permit if one is required.

2. Excavate for foundation.

3. Pour footings, lay concrete block foundation walls, bolt bedplates, build basement partitions, sill beams, girder, floor joists, subflooring.

4. Build and raise wall frames; partitions, ceiling joists, raise ridge, rafters, gable studs. Sheath walls and roof, install window and door frames. Apply #15 felt and roofing. Nail fascia trim and corner boards. Apply #15 felt and exterior siding to gable ends. Apply siding, clapboard or shingles to sides, front and back. Build bathroom partition.

5. Rough-in wiring, plumbing and heating.

6. Install insulation, apply paneling to walls, gypsum board to ceilings.

7. Install finished flooring, run base, shoe and ceiling molding, window and door trim.

8. Complete plumbing, wiring, heating.

9. Paint, wallpaper or panel.

The Easi-Bild method of contruction covers foundation, framing, flooring, sheathing and general interior work. No attempt is made to explain plumbing, wiring or installation of heating equipment. Building codes in many areas require this work be done by licensed workmen. If you want to rough-in waste, supply and electrical lines, install bathroom fixtures, build kitchen cabinets, read Books #675 Plumbing Repairs Simplified, #682 How to Add an Extra Bathroom, #694 Electrical Repairs Simplified, #758 How to Modernize a Kitchen.

SITE SELECTION

Spare the trees when selecting a site. Don't build too close to a small tree. The years fly by. Small shrubs and trees grow fast. If in doubt as to how fast a tree grows, ask a local nurseryman.

Locate house so it complements the site. Also consider the future of the neighborhood. How will it look when adjacent property is built up?

Consider the climate and how it will affect living conditions. Will your largest windows face directly into the bright summer sun? Can the low winter sun shine into those rooms where you can enjoy it the most? Will prevailing winds pile snow into areas selected for a garage or driveway? A neighbor with a garage facing direction you plan yours can relate problems he's had with drifting snow. When planted in the proper location, a small hedge of evergreens will grow into a strong snow fence. Consider these facts when selecting a site.

After you have chosen the exact site, make a plot plan, Illus. 10, showing location of house and/or garage. Indicate distance from property lines.

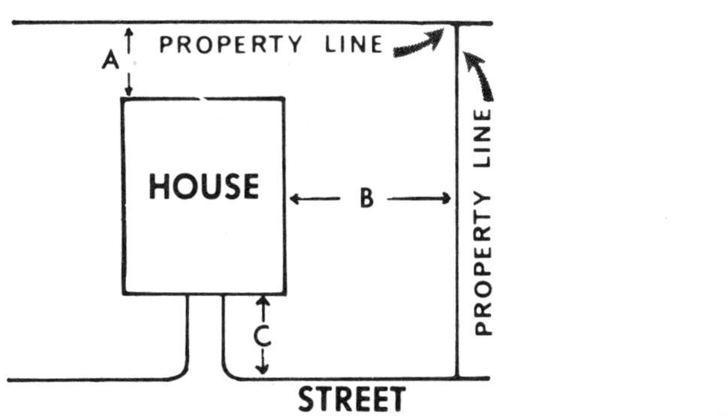

Note distance to property lines-A, B, C

(10)

FRONT ELEVATION

11

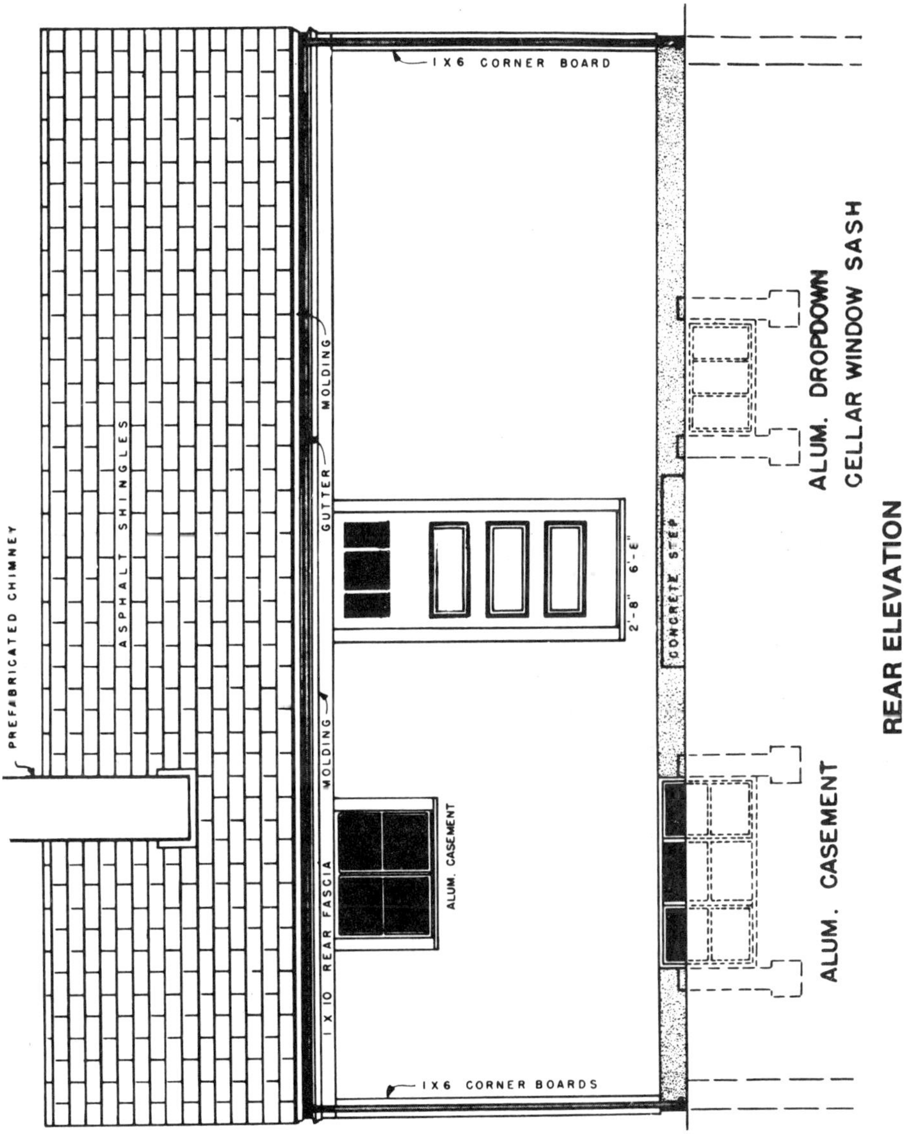

12

If a building permit is required, you will have to furnish a plot plan and elevations, Illus. 11,12,13,14.

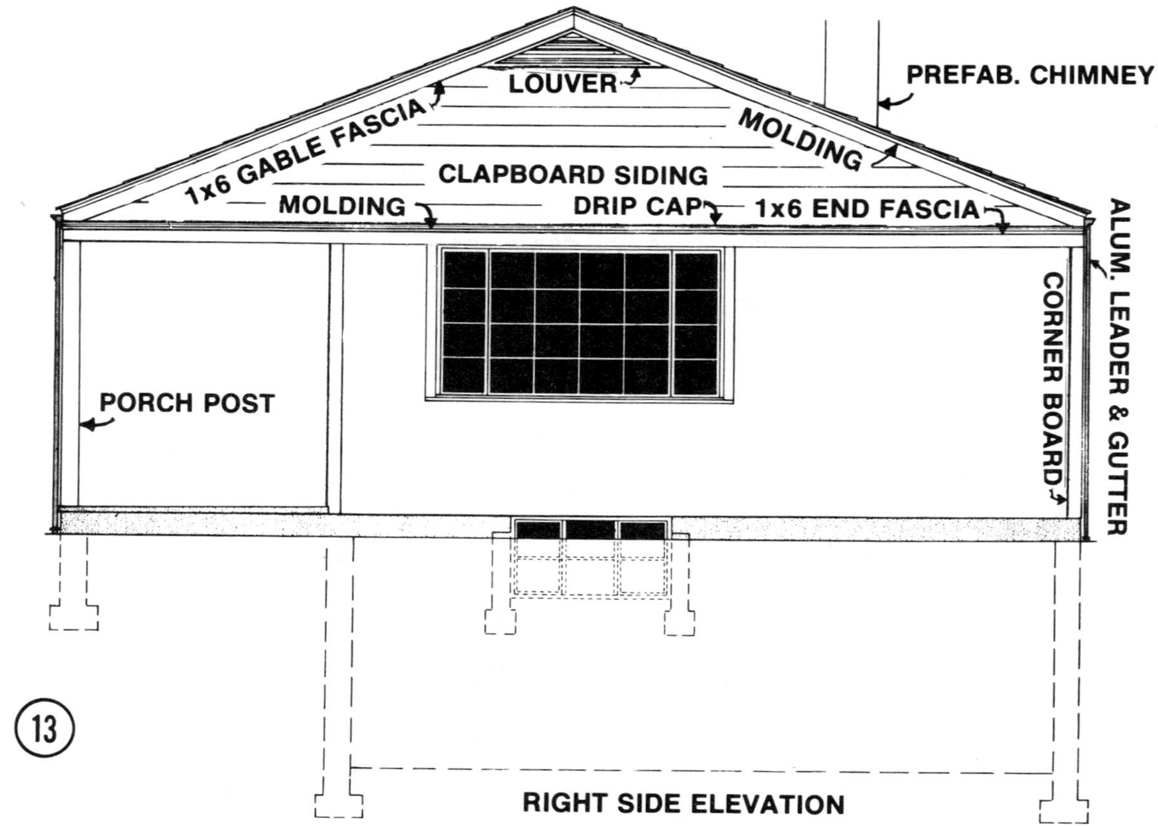

RIGHT SIDE ELEVATION

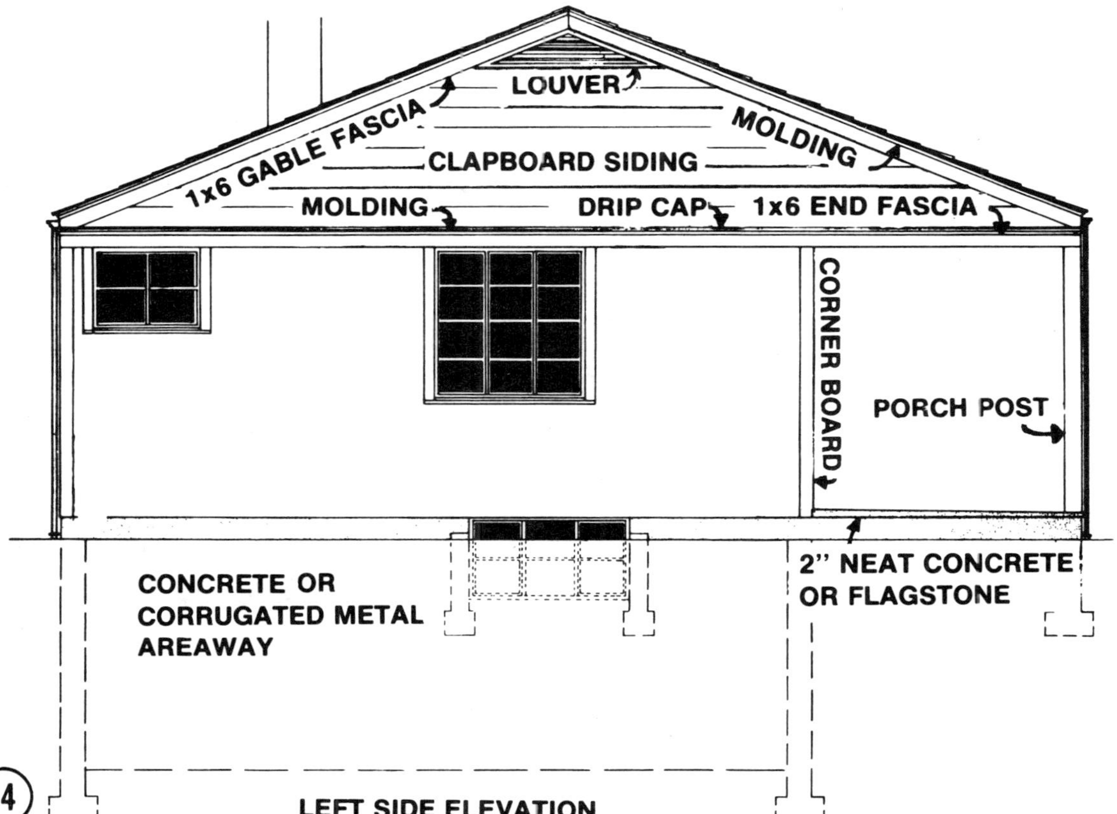

(14) LEFT SIDE ELEVATION

LIST OF MATERIAL

Foundation: Exact amount of cement, sand, gravel or total number of concrete blocks depends upon depth and thickness of footings and height of foundation wall. Buy cement and lime as you go along. Lumber varies in size. A 2 x 4 measures 1½ x 3½", 2 x 6 - 1½ x 5½", 2 x 8 - 1½ x 7¼". Rough cut lumber can measure 1⅝to 1⅞ x 3⅝ to 3⅞" for a 2 x 4. Always build frames to overall size required. Always check overall length of all inner framing members prior to cutting.

10 bags of Cement
5 bags of Mason's Lime
5 yds. Sand
5 yds. ¾" run-of-bank gravel
66 — 8 x 8 x 16" Concrete Blocks per course for house foundation (20'8" x 24'8")
29 — 8 x 8 x 16" Concrete Blocks per course for porch foundation (7'4½" x 24'8")
19 — ½ x 12" Machine Bolts, Nuts and 36 Washers
7 — ½ x 8" Machine Bolts, Nuts and 14 Washers
120 lineal feet 12" Aluminum or Copper Flashing
4 — 12" lengths of 1" Galvanized Pipe
70 feet Drain Tile
2 x 3 — 29/10
2 x 4 — 2/20,4/18,18/16,98/14,37/12,25/10
2 x 6 — 55/16,23/14,3/12,2/10,3/8
2 x 8 — 6/16,9/14,26/12,19/10
2 x 10 — 3/10
1 x 2 — 34 lineal feet
1 x 3 Furring — 400 lineal feet
1 x 4 — 4/16,2/14,1/10
1 x 6 — 16/16,3/14
1 x 10 — 2/14,2/12
¾ x 1⅝" Drip Cap — 2/16,2/14
¾ x 1¾" Molding — 8/16,6/14,2/12
13/ 16 x 2¼" Oak Flooring to cover 700 sq.ft.

3/8" x 4 x 8' Plywood — 32 panels
5/8" x 4 x 8' Plywood — 52 panels
3/8" x 4 x 8' Exterior Grade Siding Plywood — 27 panels
1/4" x 4 x 8' Hardwood Plywood — 27 panels
7/8" x 4 x 8' Plastic Laminate for Kitchen Countertops
1/16" Plastic Laminate for Kitchen Backsplash
3/4 x 8" Red Cedar Siding to cover 170 sq.ft.
1 bundle 16 or 18" Cedar Shingles
Roofing Shingles to cover 750 sq.ft.
#15 Roofing Felt to cover 2825 sq.ft.
Gypsum Board to cover 650 sq.ft.
Ceiling Tiles to cover 875 sq.ft.
Sheetrock or equal to cover 235 sq.ft.
1/4" x 4 x 8' Asbestos Cement Board — 5 panels
Finished Stairs:
9 — 10½" Oak Treads
12 — 8" Pine Risers
36 lineal feet ½ x ¾" Oak Cove Molding
36 lineal feet ½ x ¾" Oak Shoe Molding
One Newel Post
One Starting Step
4 ft. Stair Rail
7 — 1⅜ x 1⅜" Pine Balusters
7 ft. Oak Nosing
One Pull Down Attic Stair Unit for 60" opening
40 lineal feet 12 to 18" strips of metal lath
Seven 2ft. and one 1 ft. length Prefabricated Chimney complete with necessary fittings
5 — 10' lengths Alum.O.G.Box Gutter - 4 sections with outlet
4 — 10' lengths Alum.2⅜x3¼" Rectangular Downspout
Attaching parts for gutters:
8 Slip Connectors, Mastic
Two left and Two Right End Caps
2 doz. Strap Hangers, Nails
4 Alum. Strainers
8 Pipe Bands
28 lineal feet 14" Copper Flashing
Aluminum Foil Insulation or equal to cover 2100 sq.ft.
Masonry Waterproofing to cover 750 sq.ft. with two coats
4 — Pull Chain Ceiling Fixtures

LIST OF MATERIAL (continued)

5 ft. Recessed Tub
Water Closet
20 x 18" Lavatory
or 1 Plumbing Wall complete with wall hung toilet, tub and lavatory.
2 — Aluminum Vari-Pitch Louvers
10 lbs. Putty
1 gallon Exterior Primer
2 gallons Exterior Paint
15 gallons Asphalt Waterproofing
2 — Folding Table Brackets
1 pr. Desk Slides
Continuous Hinge — 5 ft. length
Built-in Range
Ventilating Fan
Combination Sink and Dishwasher
Eye Level Oven
Colonial Black Ornamental Hardware for Kitchen Cabinets:
  4" Hinges
  Door Pulls
  Door Catches
2 pr. 4 x 4" Loose Pin Butt Hinges
1 pr. 3½ x 3½" Loose Pin Butt Hinges
5 pr. H-Hinges, 7" long
5 pr. ½" Latch Sets
One 36" and One 48" Sliding Closet Door Unit
Corrugated Iron Areaways to fit basement windows or
  4 x 8 x 16" concrete blocks to build areaways
DOORS: All doors complete with frames, trim, saddles
  1 — 2'8" x 6'8" Dutch Door (exterior)
  1 — 2'8" x 6'8" Exterior Door
  2 — 2'0" x 6'8" Flush Panel Door (inside)
3 — 2'6" x 6'8" Flush Panel Doors (inside)
Locks and necessary Dutch Door Hardware
WINDOWS:
  8 — Aluminum Casements
  1 — Drop-down Cellar Window Sash

NAILS:

1 keg 8d Common Nails
1 keg 16d Common Nails
1 keg 6d Common Nails
50 lbs. big head Roofing Nails
5 lbs. 6d Aluminum Nails
15 lbs. 4d Galv. Finishing Nails
15 lbs. 8d Steel Cut nails
5 lbs. 4d Finishing Nails
15 lbs. 6d Finishing Nails
15 lbs. 8d Finishing Nails
5 lbs. 3d Blue Lath Nails
10 lbs. 8d Cut Nails
1 lb. each ¾, 1 and 1¼" Wire Brads

Recommended:

#605 How to Apply Paneling
#682 How to Add an Extra Bathroom
#696 Roofing Simplified
#758 How to Modernize a Kitchen
#816 How to Lay Ceramic & Quarry Tile
#675 Plumbing Repairs Simplified
#694 Electrical Repairs Simplified
#697 Forms, Footings, Foundations, Framing, Stair Building

LAY OUT GUIDE LINES

The first step is to decide what grade level of ground you want to establish around the completed house. Drive a stake flush into ground at grade level selected. Drive a nail into top of this stake, allowing it to project about 1", Illus. 15. Tie a length of line to the nail and stretch line to various points within the area of the proposed building. Place a line level on line. Note low and high points. This will give you some idea where to remove soil, where to fill in, where to bury a fuel oil tank, etc.

Next drive a stake into ground to indicate a front corner. This stake can project 3 to 4" above ground. Drive a nail into top of this stake. Approximately 3' from this corner, drive a batter board stake. Use 1 x 4 sharpened at one end. Batter board stakes can project above batterboards, or can be sawed off flush with top edge of batter board, Illus. 15,16.

Stretch a line from the grade level stake to batter board stake. Attach level to line. When line indicates level, mark stake. This mark now indicates "grade level."

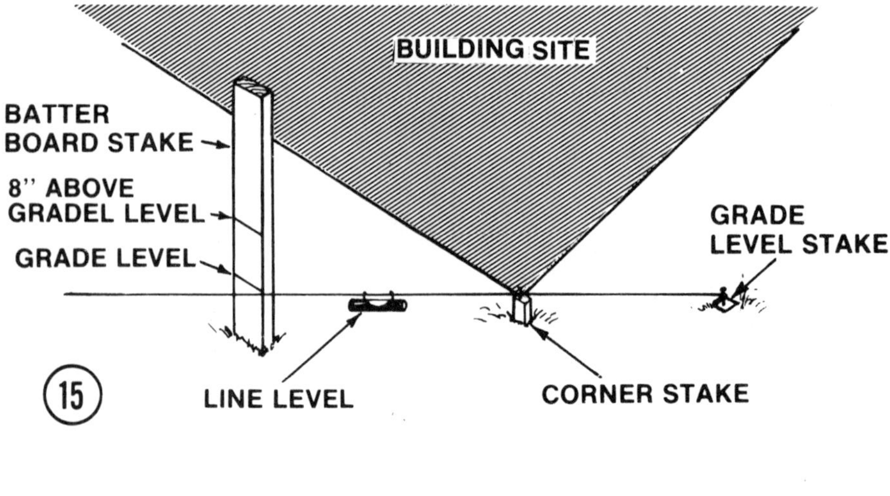

(15)

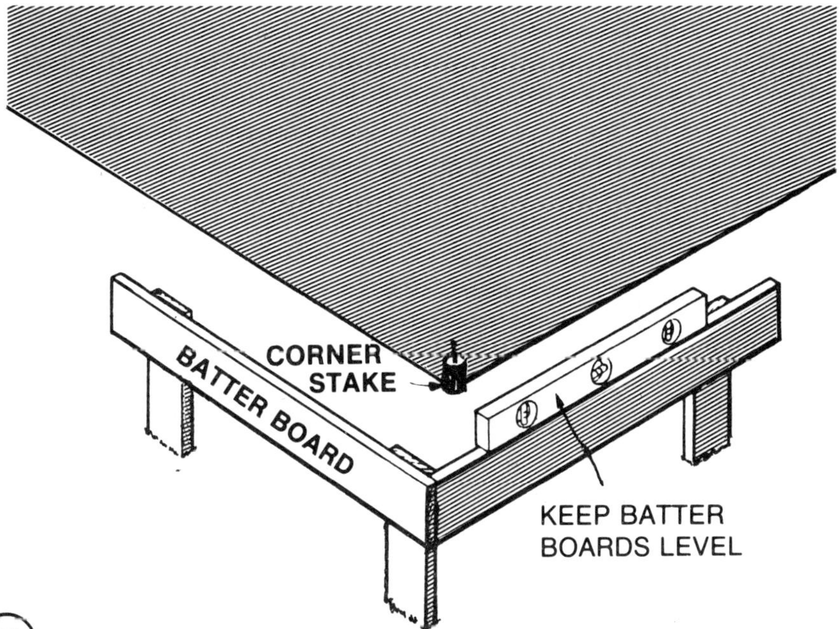

16

Top edge of batter board represents height of finished foundation wall.

Measure up 8'' if you want one course of block to show above grade; 16'' for two courses; 24'' for three courses, etc. Nail batter boards to stake at height above grade selected. The top edge of batter board now indicates top of finished foundation wall. Foundation should not be less than 8'' above grade. Wood frame should be placed 8'' or higher above grade to eliminate contact with earth.

Nail batter boards to stakes at height required. Remember, top edge of batter board should be level and at height that indicates top of foundation wall.

Overall foundation dimensions show how to build a minimum size house. Alter these in length or width to suit your needs. Include the porch as part of the living and bedroom if you like, Illus. 6.

To establish front building line, Illus. 17, hook end of a steel measuring tape on nail in front corner stake and measure 24'8''. Drive a stake into ground, drive a nail into this stake to indicate 24'8''. Be sure to allow 24'8'' from nail to nail. Three feet away from this corner stake, erect batterboard stakes. Be sure front building line is parallel to, or at right angle to property line, and distance codes require.

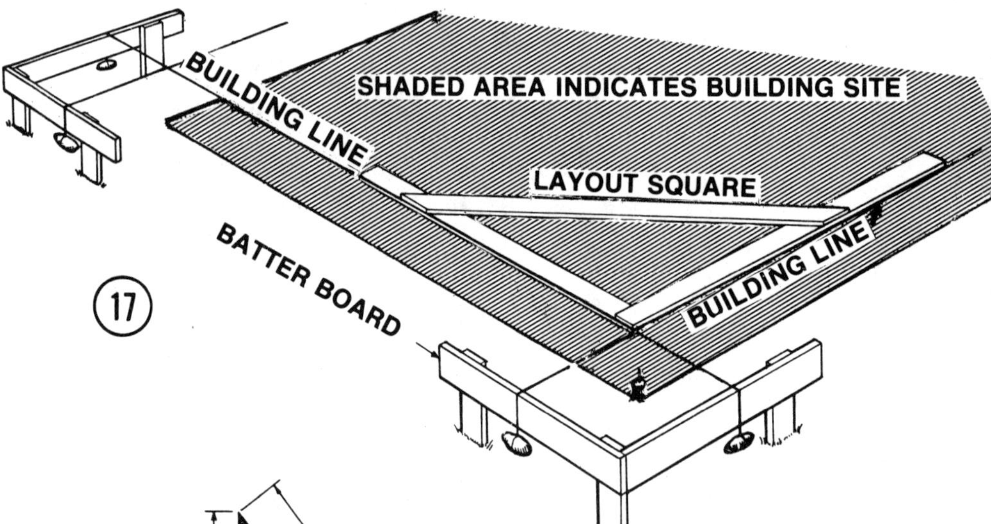

17

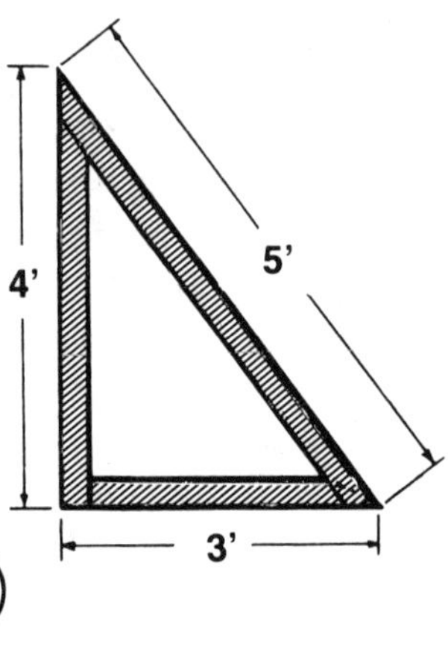

18

Make a layout square, Illus. 18. To make a layout square, square off ends of 1 x 4. Mark three feet on one board; four feet on another; five feet on a third. Nail or screw boards together as shown.

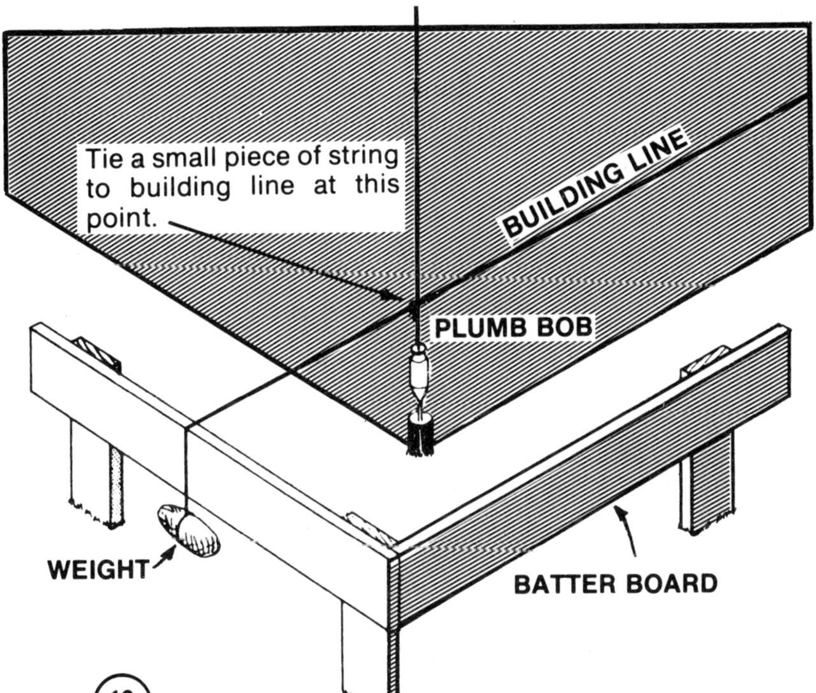

Place layout square in one corner, Illus. 17. Sight down square. With plumb bob to guide you, Illus. 19, stretch guide lines above nail that indicates exact corner. To hold taut, lines can be tied to a brick or stone. The line now indicates both height and outside edge of completed foundation block wall. Check line with line level. Make saw cuts in batter board to lower line. Drive a nail in batter board if you need to raise the line.

Tie a piece of building line, Illus. 19, to indicate corner. To make certain lines are square, measure diagonals, Illus. 20. Lines are considered square when diagonals measure equal length.

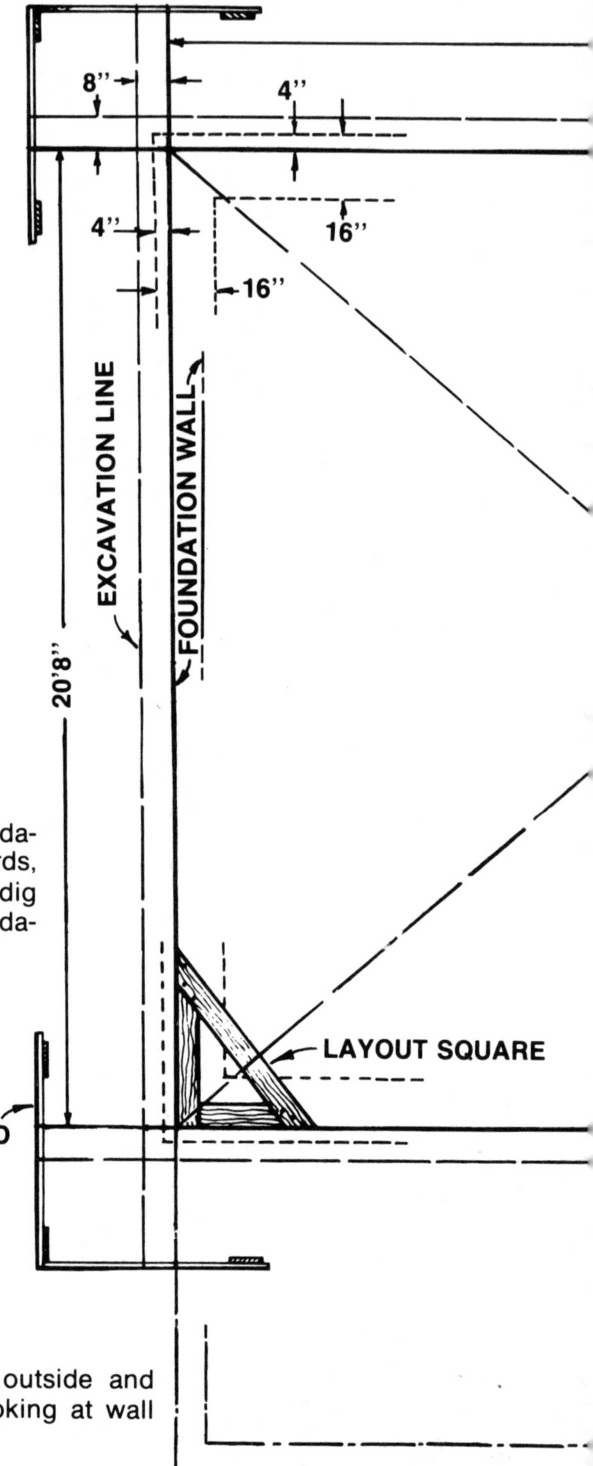

NOTE: To build porch foundation, erect batter boards, stretch guide lines and dig trenches after main foundation is completed.

20

Consider yourself on the outside and facing the house when looking at wall frames.

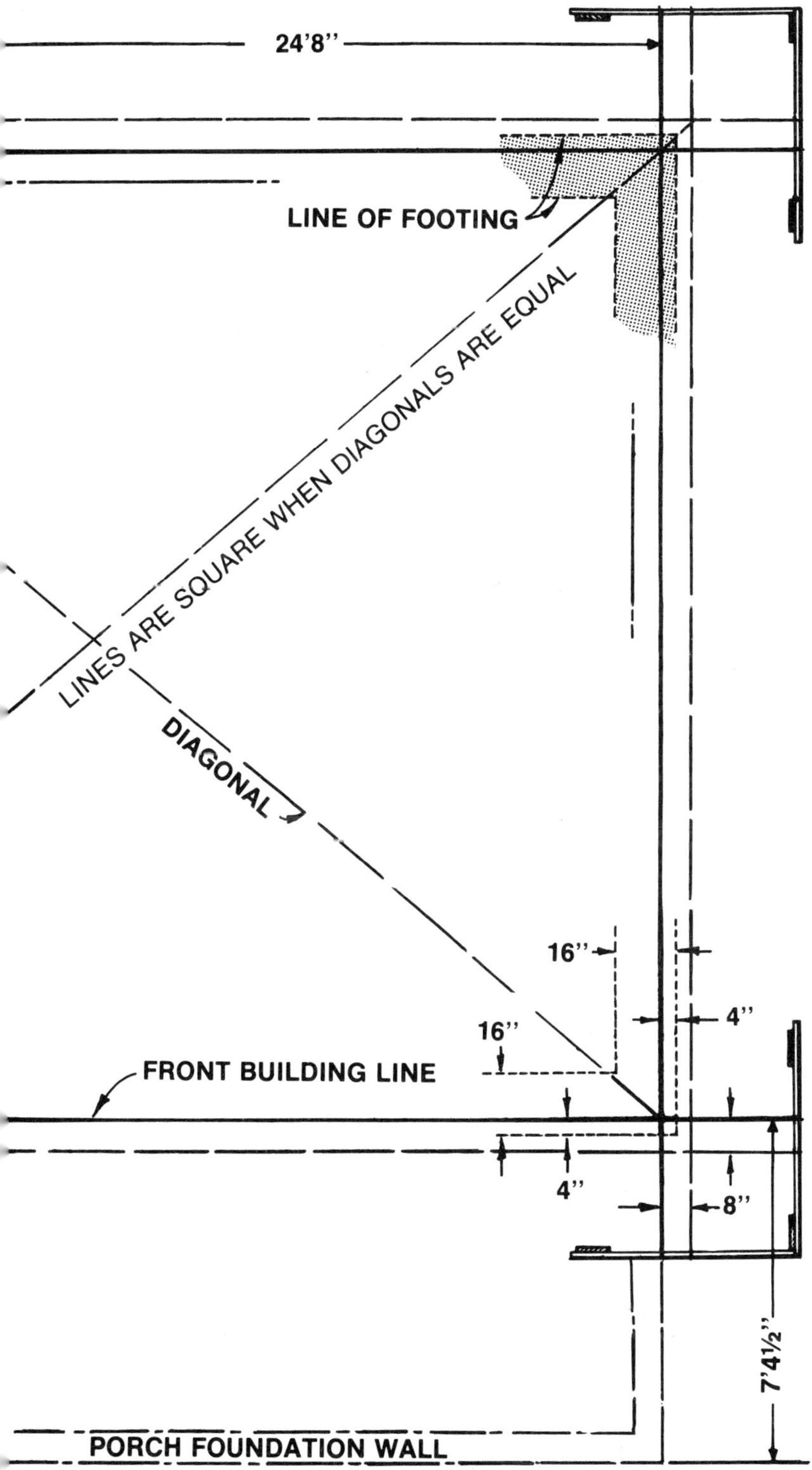
24'8"
LINE OF FOOTING
LINES ARE SQUARE WHEN DIAGONALS ARE EQUAL
DIAGONAL
16"
16"
4"
FRONT BUILDING LINE
4"
8"
7'4½"
PORCH FOUNDATION WALL

FOOTINGS

Using lines as a guide, excavate to depth required for a crawl space or basement. Remove top soil and pile it far enough away from site selected so it doesn't get in the way of storing and handling materials during construction. Top soil is valuable. Save all you can.

Trenches for footing must be dug to depth required. Footings must be laid on undisturbed soil. Footings for 8" blocks should be 16" wide, 3½ to 5½" thick.

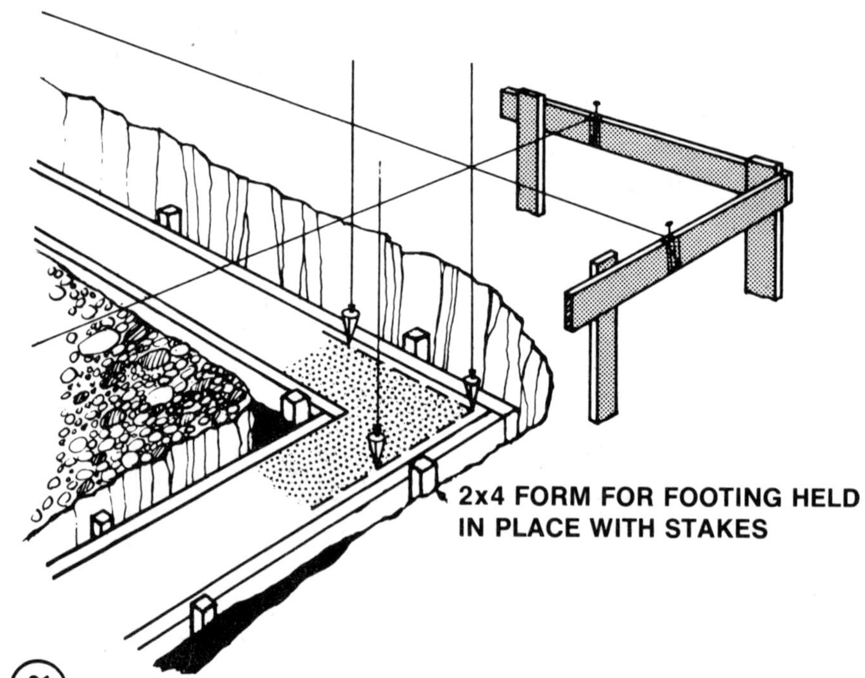

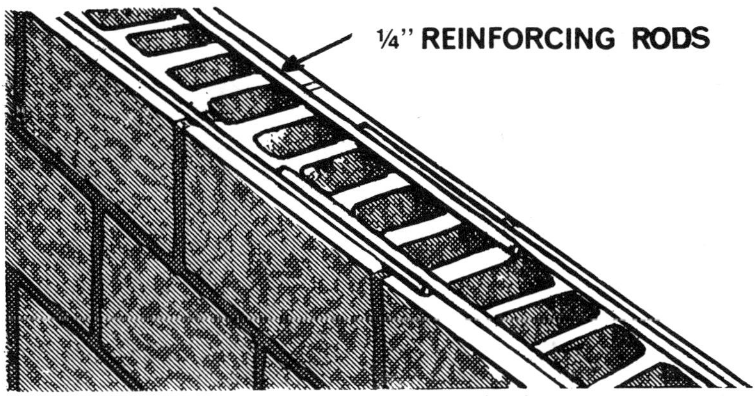

(22)

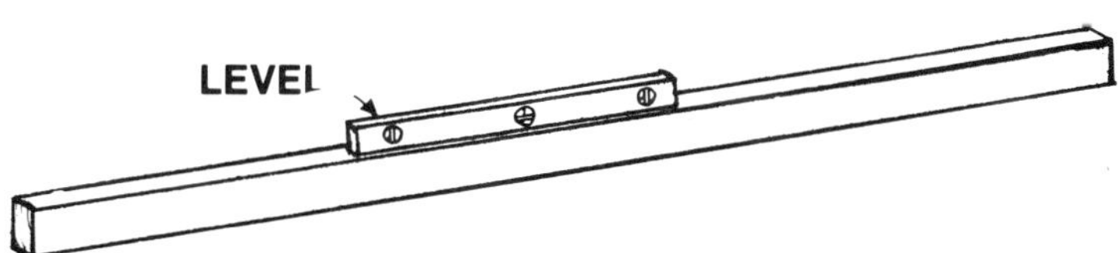

(23)

Place 2 x 4 or 2 x 6 forms for footings, Illus. 21. Estimate distance from top of form to guide lines so it can be divided by eight for 8" concrete block. If you are building a block wall against a slope where there's any chance of pressure building up, use 10 or 12 x 8 x 16" blocks instead of 8 x 8 x 16. Use reinforcing wire or rod your dealer recommends between each course, Illus. 22.

To make certain bottom of trench is level and exact distance from guide lines, use a 2 x 4, 2 x 6 or 5/4 x 6 x 8' edgewise, with a carpenter's level, Illus. 23, to test. Be sure lumber selected has a straight edge.

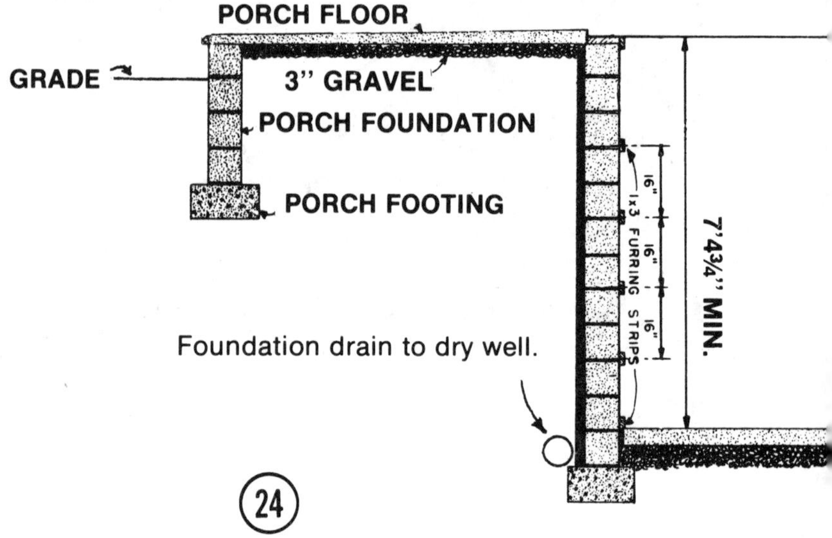

BASIC HOUSE
Foundation measures 24'8"x20'8".
Overall framing measures 24'7¼"x20'7¼"

If you don't plan on a full basement, excavate at least 48" beneath house to provide free circulation of air, plus crawl space for running pipe, heating, wiring, etc. Consult your plumber and electrician to make certain they will rough-in in this amount of space.

All exposed water lines in a crawl space should be insulated. If you heat with hot air, insulate ducts.

Many heating contractors recommend digging a small area deep enough in a crawl space to install a hot air furnace. Others install by hanging it from floor joists. Some use an insulated crawl space as a plenum. The furnace heats

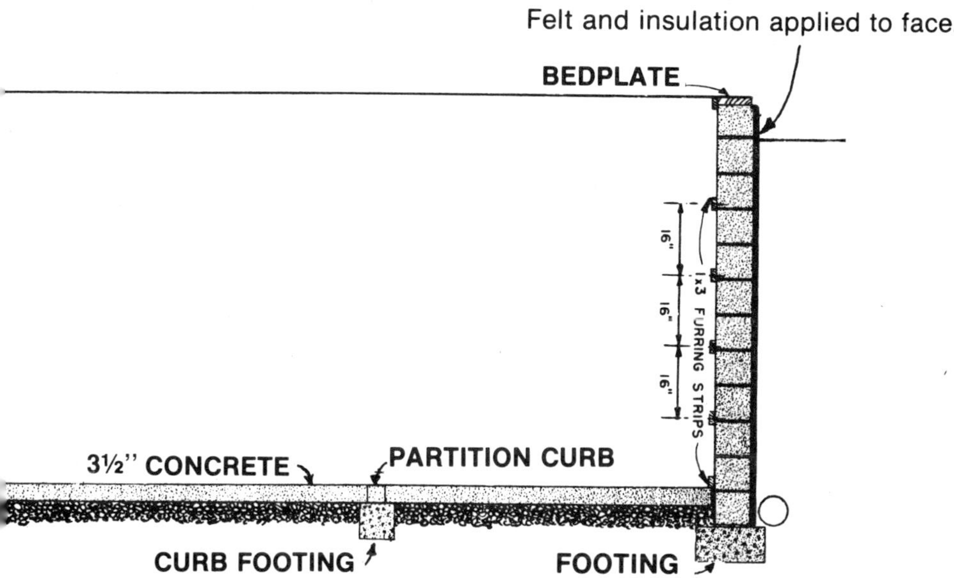

Add an extra course of block if you want 8" more headroom. You can also lay gravel level with top of footing and lay concrete floor on shoulder of footing.

ALLOW A ⅜ TO ½" EXPANSION JOINT BETWEEN FLOOR AND BLOCK WALL. SEAL JOINT WITH ASPHALT.

plenum. Hot air is distributed to each room through floor or wall registers. If you use crawl space as a plenum, lay a concrete floor in crawl space to make area as dust free as possible. Apply felt and insulate outside face of block, Illus. 24.

Plumbers frequently want to start roughing-in, in a crawl space, after floor joists are in place, and before subflooring has been completed. If they do it later, they must enter crawl space either through a window placed in top three to four courses of block, or through a trap door cut in floor. A 16 x 30" trap door enables some plumbers to go below; a 32 x 30" trap door is more convenient.

EXCAVATE FOR FULL BASEMENT

Illus. 24 shows footings and foundation walls required for basement having 7' plus headroom. Illus. 25 shows how first floor and basement can be built as a side hill or earth covered shelter.

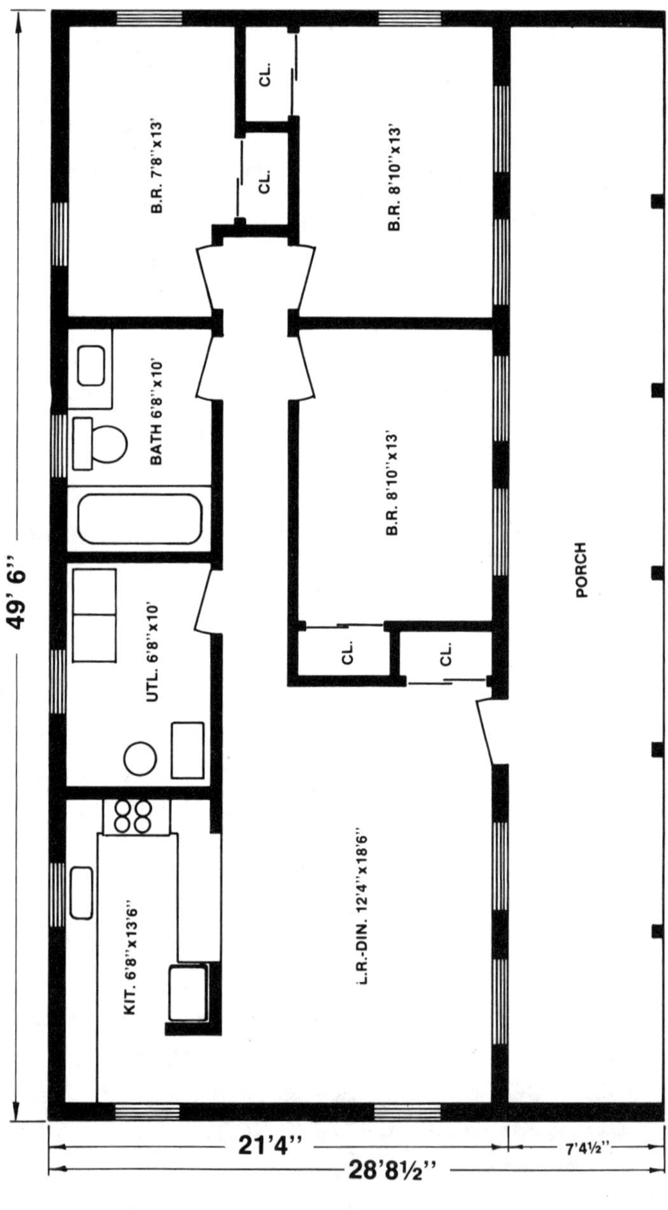

25

Illus. 26 shows footings, foundation walls plus footings for concrete block partitions. Follow same procedure for an expanded foundation or side hill.

Excavate approximately 8' 0'' to 8' 6½'' below guide lines for a minimum height basement. This allows 3'' of gravel, 3½'' for footings. Dig to depth required if you want more height in basement or more foundation block above grade.

Dig trenches for footings to depth indicated or depth required for ceiling height local codes specify, Illus. 9,24.

If soil is firm, trenches for footings can be dug width required. Concrete can be poured in trenches. If you need forms, use 2 x 4, 2 x 6, 2 x 8 or width lumber required. Drive stakes in ground to hold forms in place.

If codes, for example, require an 8'' thick footing, trench for footing should be 8'8'' from guide line. This will permit 8'' footing and 12 courses of block. Each course, including mortar, is figured at 8''.

Use one part cement, three parts sand to five parts gravel for footings, or buy ready-mix. Allow footings to set three days before laying concrete block. If you build during freezing weather, use Antro-Hydroset or equal anti-freeze in concrete. Always cover concrete at night to protect against freezing.

To obtain the best price on sand and gravel, order in 5 to 7 yard lots. Buy cement and lime as you go along. Don't store quantities on the job. Dampness may get in and harden it.

Talk to the plumber, heating contractor and electrician to find out when each wants to start roughing-in. Follow their recommendation concerning where water, fuel oil, gas, waste, electric and telephone lines go through footings or foundation. Ask the electrician about running an underground service cable to the house.

ALTERNATE METHOD OF CONSTRUCTION

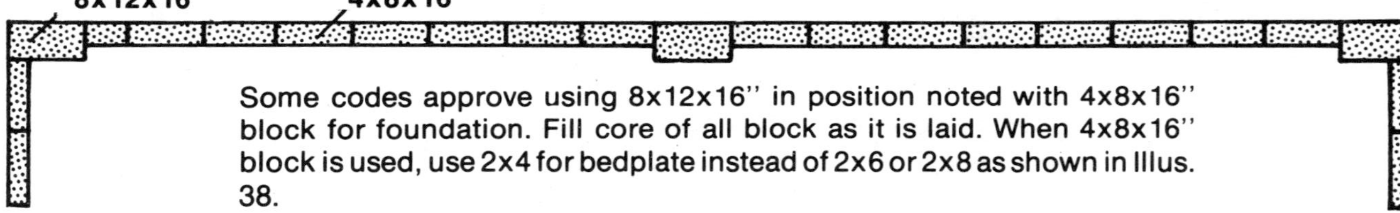

Some codes approve using 8x12x16" in position noted with 4x8x16" block for foundation. Fill core of all block as it is laid. When 4x8x16" block is used, use 2x4 for bedplate instead of 2x6 or 2x8 as shown in Illus. 38.

26

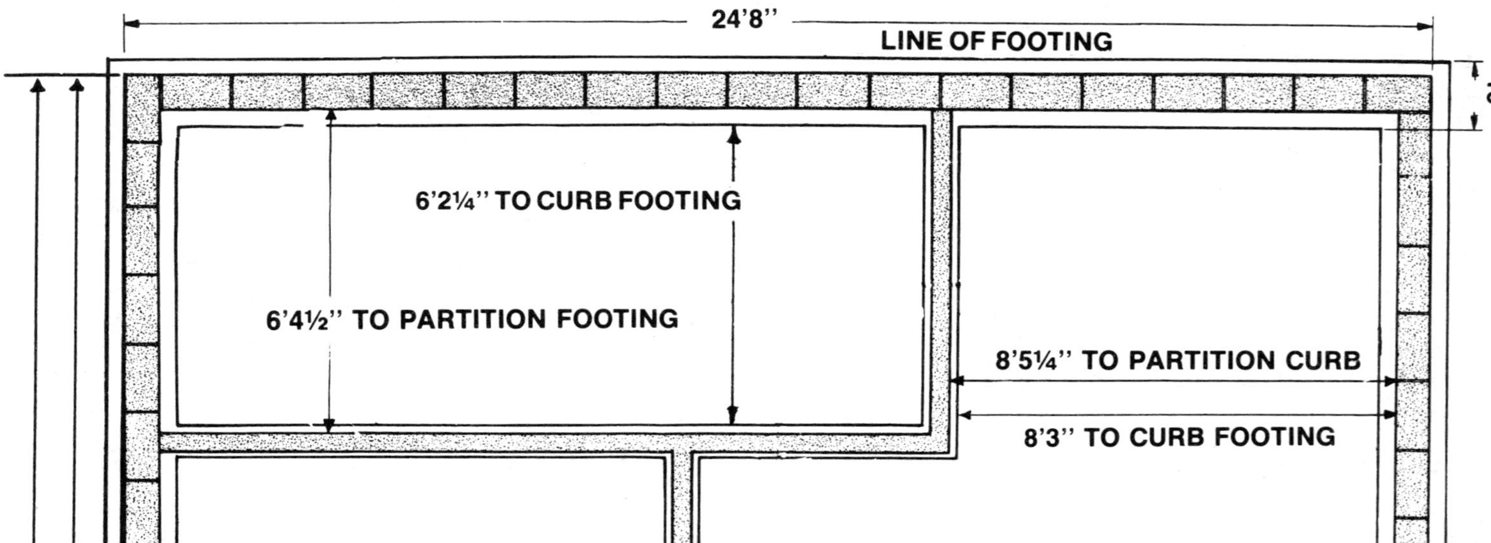

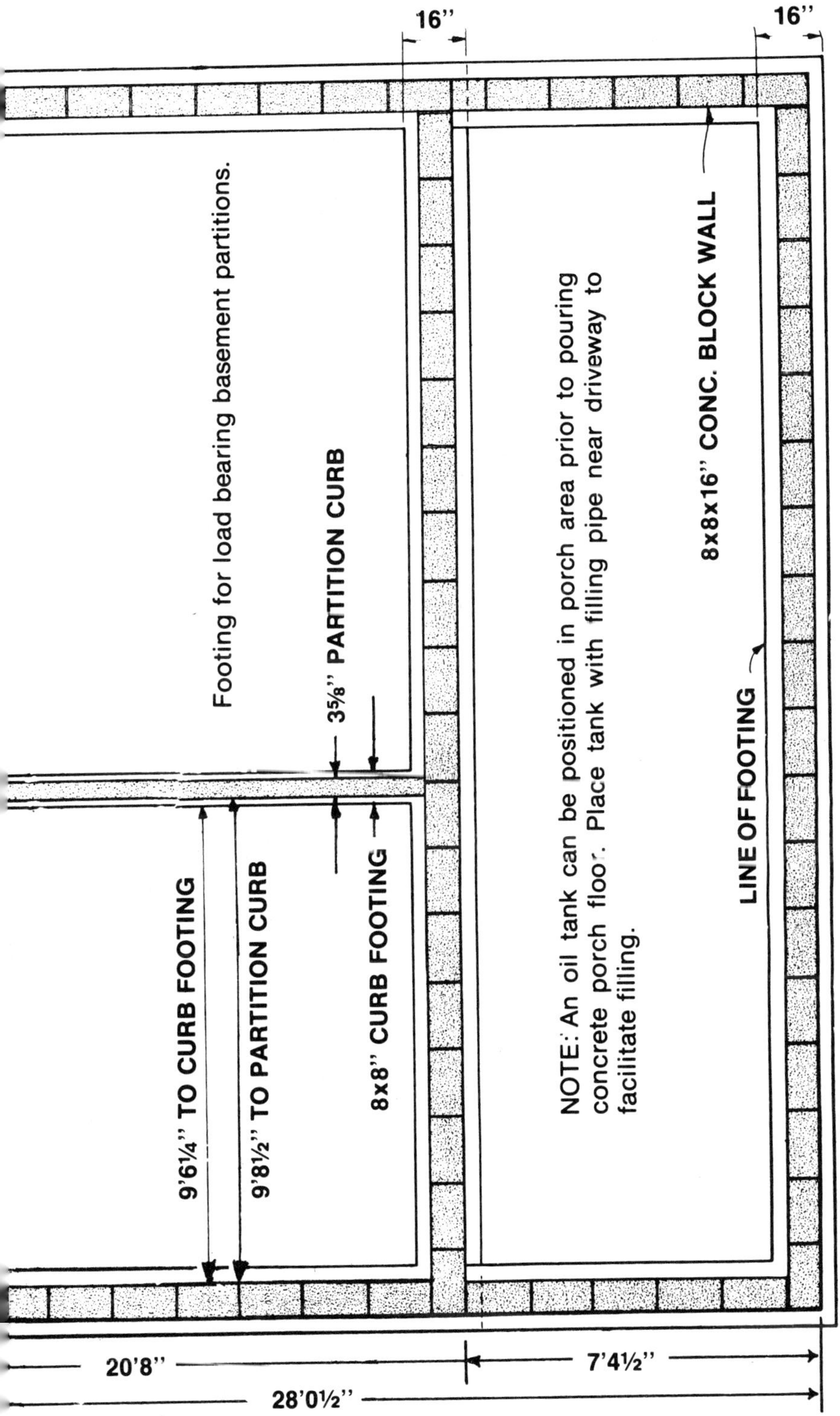
16"
16"
Footing for load bearing basement partitions.
3⅝" PARTITION CURB
8x8" CURB FOOTING
9'6¼" TO CURB FOOTING
9'8½" TO PARTITION CURB
NOTE: An oil tank can be positioned in porch area prior to pouring concrete porch floor. Place tank with filling pipe near driveway to facilitate filling.
8x8x16" CONC. BLOCK WALL
LINE OF FOOTING
20'8"
7'4½"
28'0½"

To provide an opening for these lines, make an open end box from four pieces of 2 x 6 or 2 x 8 x 16'' if lines enter through footings; 8'' long if lines enter through foundation wall, Illus. 27, or you can use 4 or 6'' diameter tile instead of 2 x 6. Place in position indicated. Mark location and depth of box on foundation plan. Remove 2 x 6 after concrete has set. Be sure to keep a plan showing location of utility lines for reference should repairs be required. Phone the telephone company to inquire whether they will provide an underground cable. This can be buried in the trench dug for water and electric line.

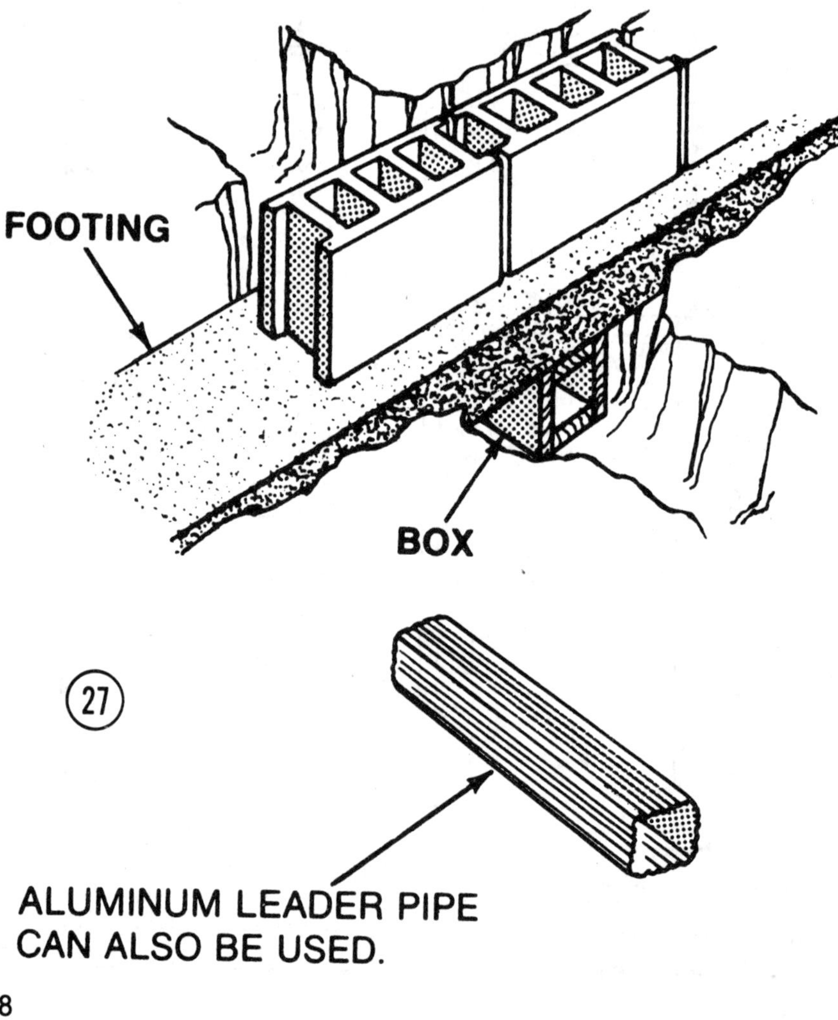

FOUNDATION WALLS

Note Illus. 26, 28.

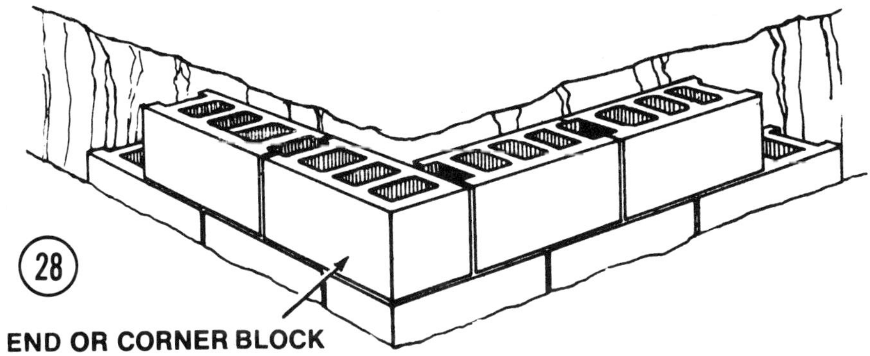

Estimate approximately 66 — 8 x 8 x 16" concrete blocks per course for house, plus 29 — 8 x 8 x 16" concrete blocks per course for porch. Buy blocks required for an extended house, Illus. 6, or for the earth sheltered house, Illus. 25.

If foundation work is done over a period of time, check building lines at the beginning of each day's work to make sure lines are level and in correct position.

Purchase extra blocks to allow for breakage. When necessary, make heavier mortar joints between blocks to fill space available. Figure three blocks (per course) to every four feet. Corner blocks, Illus. 28, can be obtained for exposed outside corners above grade.

Before laying blocks, check guide lines to make certain they are square, level and taut. Check diagonals. Setting blocks takes some skill, common sense and practice. You learn much by working alongside an experienced mason. Most masons use a mortar consisting of 1 bag of Portland Cement, ⅓ to ½ bag of mason's lime to 20 to 24 shovels of screened sand. Brixment, a cement specially prepared for use on blocks or brick, can also be used. It requires no lime. Follow manufacturer's directions.

A skilled mason uses mortar on the firm side, not too wet, and not too dry. Watch one work and notice the consistency he uses. To build foundation wall, start by spreading a thick layer of mortar on footing at one corner.

To locate exact corner, drop a plumb line down from guide line, allowing bob to mark mortar. Drop bob again about 2' from corner on each side. With straight edge and trowel, draw line in mortar to indicate exact corner. Set a corner block level and square along drawn lines. Tap block into place. Check with level. Do the same at other end of wall. Stretch a guide line, Illus. 30, between corner blocks.

Butter up end of next block, Illus. 29, place and tap into position. Check with level. Check with straight edge.

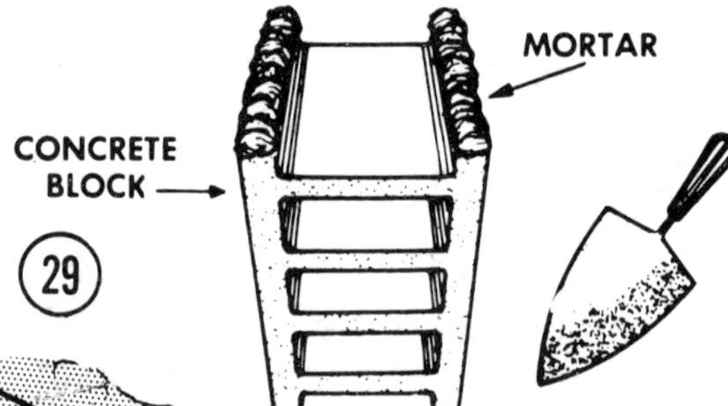

29

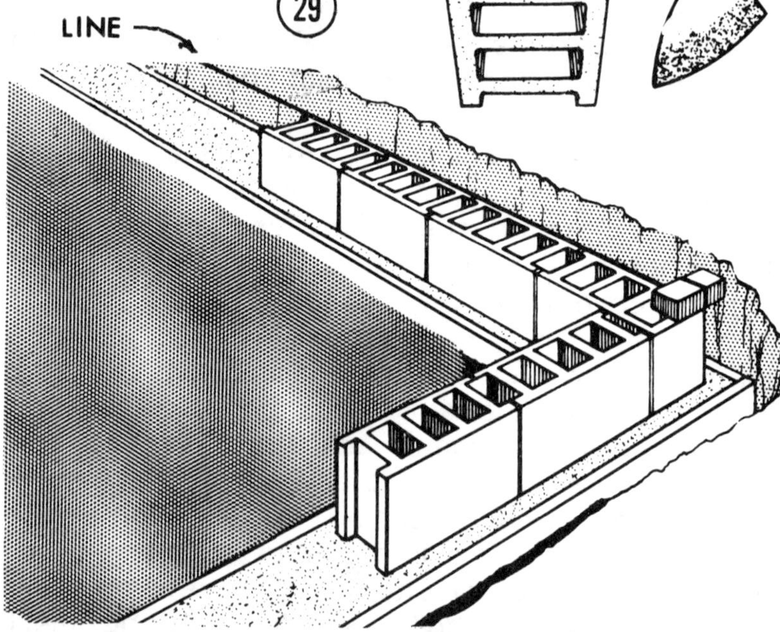

30

Use end blocks on each course above grade.

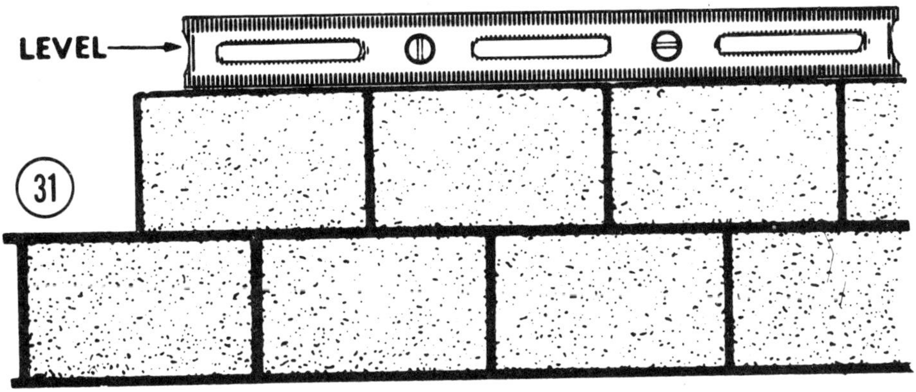

31

Stretch a guide line along top outside edge of corner blocks, Illus. 30. Lay each course to this line. Be sure to overlap joints on previous course, Illus. 31.

Always raise guide line for each course of block, Illus. 30. Be sure to overlap joints on previous course, Illus. 31.

Don't lay concrete block in freezing weather unless anti-freeze recommended by concrete manufacturer is mixed with water. Protect mortar against freezing.

If there's back pressure from a hill, fill all cores in blocks with concrete. If the pressure is stronger than normal, your materials dealer can supply reinforcing rod or wire and explain how to use.

Be sure to fill top course of block with concrete when setting anchor bolts.

Illus. 32,12,13,14 indicate position and opening for windows in foundation walls. Alter opening to size your dealer recommends for steel or aluminum windows he has available. Build block window wells into foundation. These can be larger than window opening. When protected in winter with clear, rigid plastic covering, it generates solar heat. Or use a galvanized form, Illus. 33.

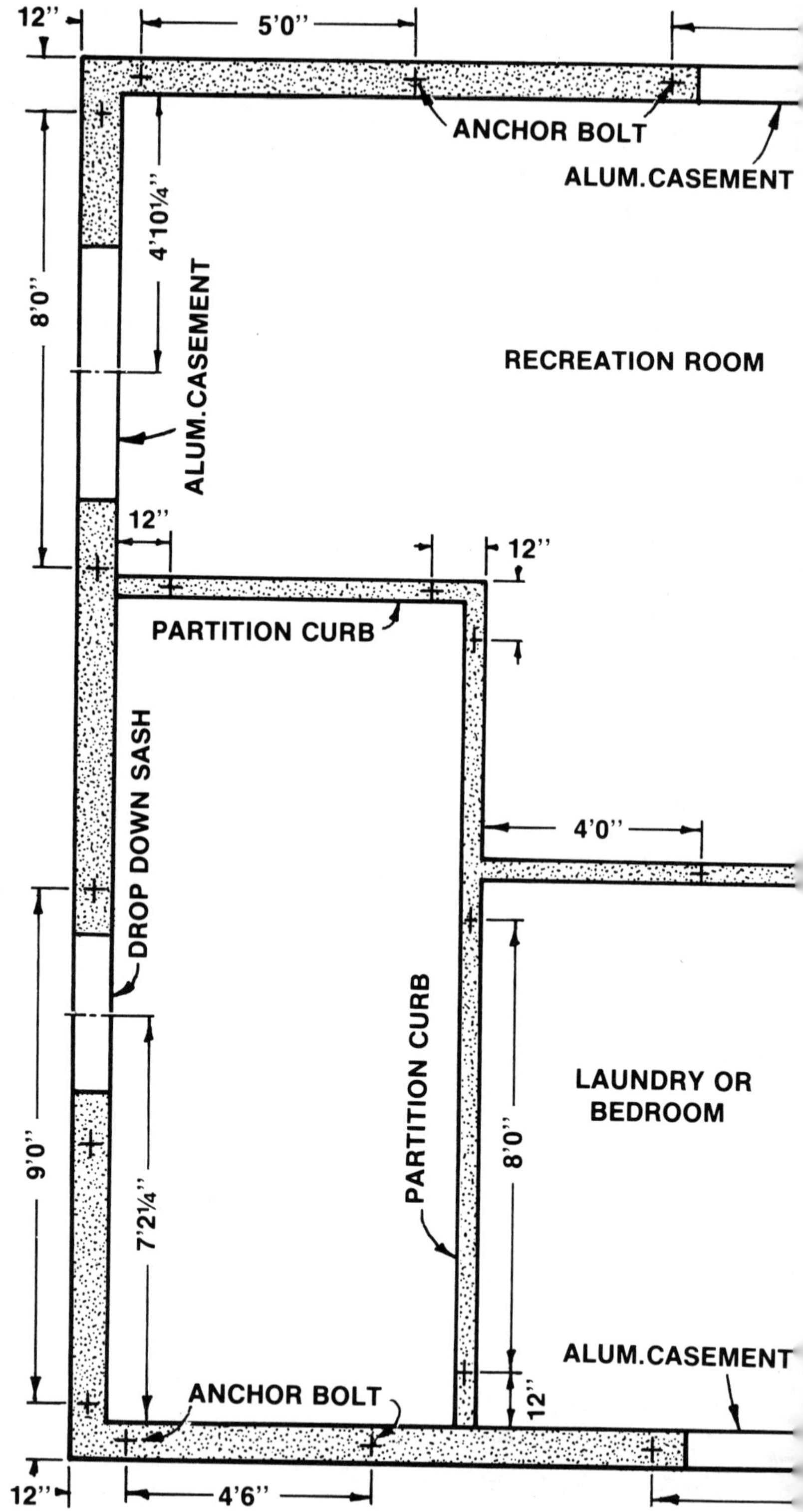
12"
5'0"
ANCHOR BOLT
ALUM.CASEMENT
4'10¼"
8'0"
ALUM.CASEMENT
RECREATION ROOM
12"
12"
PARTITION CURB
DROP DOWN SASH
4'0"
PARTITION CURB
LAUNDRY OR BEDROOM
9'0"
7'2¼"
8'0"
ALUM.CASEMENT
12"
ANCHOR BOLT
12"
4'6"

32

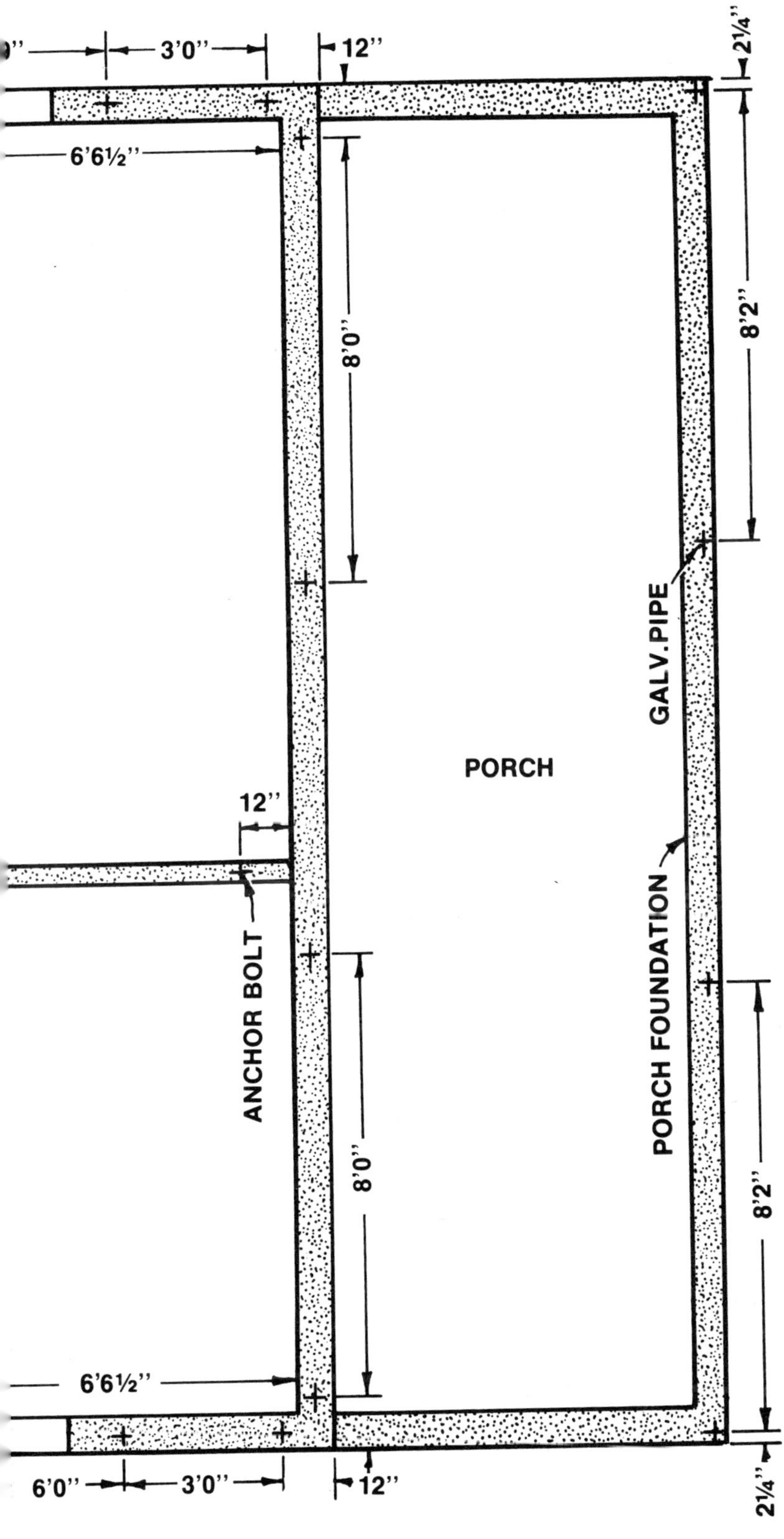
3'0"
12"
2¼"
6'6½"
8'0"
8'2"
GALV. PIPE
PORCH
12"
ANCHOR BOLT
PORCH FOUNDATION
8'0"
8'2"
6'6½"
6'0"
3'0"
12"
2¼"

Shop for aluminum, steel or wood windows recommended for installation in a concrete block foundation. Since these are frequently "stoned" by a power mower, glazing with plastic panels is recommended.

Use drop down or sliding cellar window sash. If necessary, size of opening can be altered. Bricks or concrete can be used to alter opening.

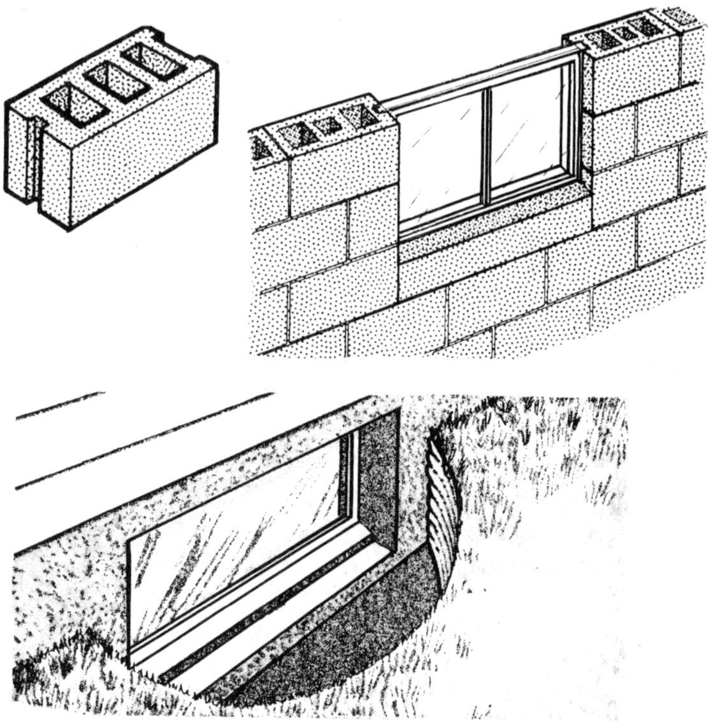

33

Concrete block with channel for metal windows is available, Illus. 33. Use these in position required to accommodate window selected.

If slotted block is used, basement window frame can be slid into position before installing bedplate. If regular block is used, basement window frame can be held in plumb position with nails driven into mortar joints. Mortar can then be applied to permanently seal frame in position. Lift out windows before setting frame. Ask dealer for installation directions for window he sells.

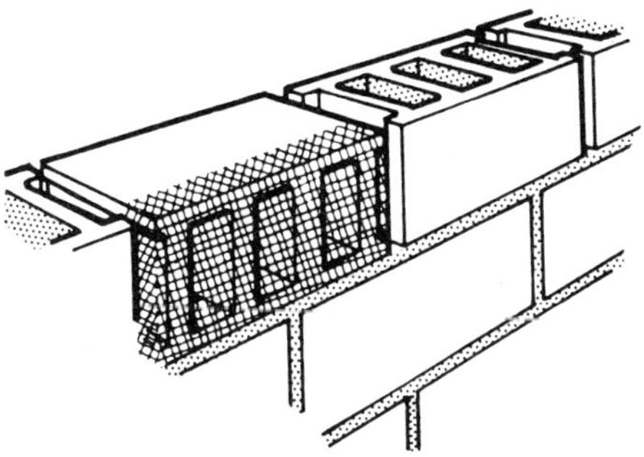

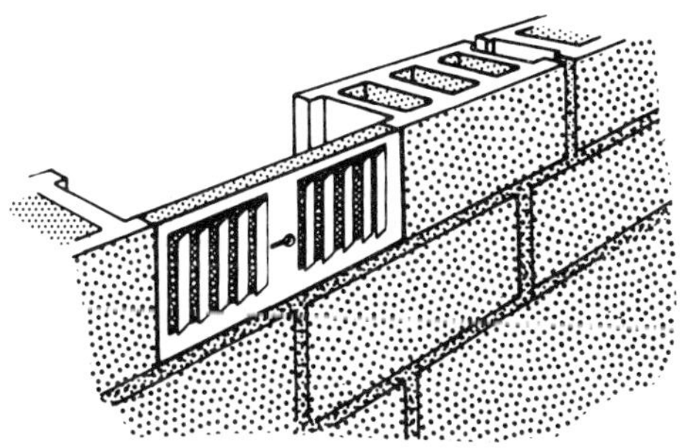

34

Always build foundation wall up to height of window. Lay blocks on both sides of opening, then continue laying block to corner. Cut block to fit.

If you build foundation with crawl space, provide ventilation with stock aluminum ventilators, or place one block on its side, Illus. 34, in each wall except front. Follow ventilator manufacturer's directions for installation. If blocks are used instead of ventilators, cover each with copper screening to keep out bees, rodents, etc.

Those instaling windows below grade should install a corrugated aluminum or steel well, Illus. 33, or build same using concrete block. Be sure to dig footings for a block wall below frost level. Fill area with 4'' of gravel or bluestone to prevent mud from splattering window.

If slope of site permits, larger windows can be installed. In this case, no outside window well is required.

SET ANCHOR BOLTS

Use 18 — ½ x 12'' machine bolts, nuts and 36 flat washers for house foundation. Seven additional ones for partition curb.

Locate and set anchor bolts 4'' in from outside edge of 8 x 8 x 16'' block. Note position indicated, Illus. 32.

Place washer at head of bolt, set bolt into block, fill core with concrete allowing threaded end to project about 4'' above block. Be sure to place all bolts in line, 4'' in from outside edge of block. Snap a chalk line on block and embed bolts along this line.

To waterproof walls, apply two ¼ to ½'' coats of cement plaster to walls. Use waterproofing when mixing one part of cement to two parts of sand. Note special instructions on page190.Trowel plaster smoothly over entire wall. Allow to cure thoroughly. Spray with a fine mist at least once a day.

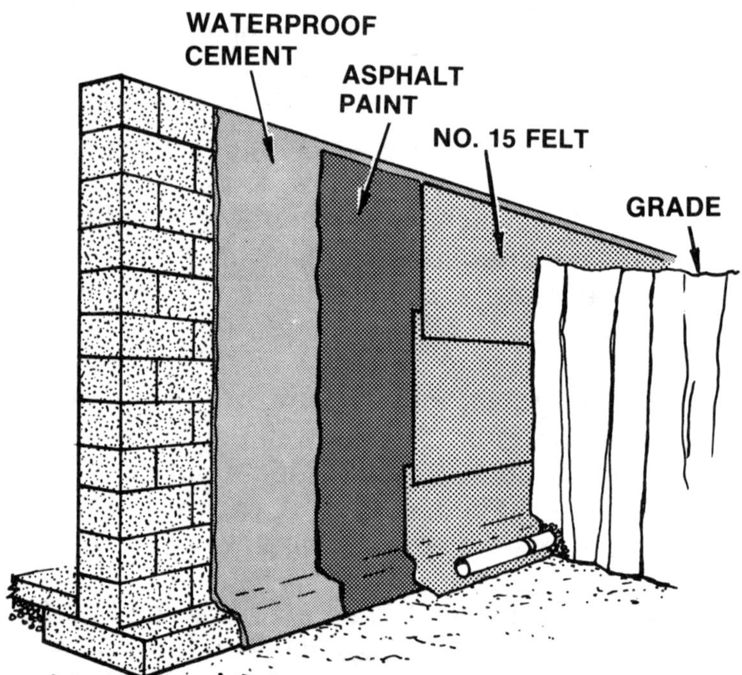

35

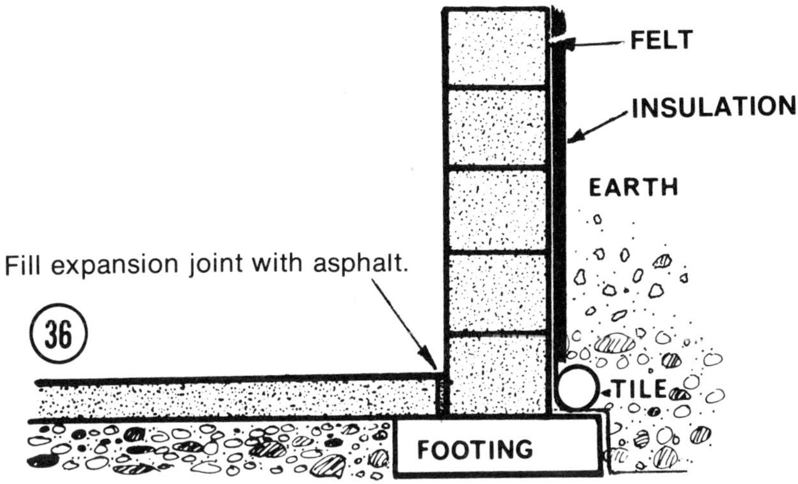

When thoroughly dry, paint with asphalt. Apply #15 felt, Illus. 35. Overlap each joint 6". Apply another coat of asphalt and another layer of #15 felt. Be sure this course overlaps joint in first course by at least 12". Rigid insulation can be bonded in position against felt, Illus. 24. Buy thickness recommended for temperature drop in your area. Cover with asphalt and bond cement asbestos board to protect insulation.

Place drain tile along outside edge of footings, Illus. 36. Slope tile to a dry well, placed some distance from foundation, Illus. 37. Wrap #15 felt around each joint of tile. Cover tile with 4" of gravel. Cover gravel with #15 felt, then backfill.

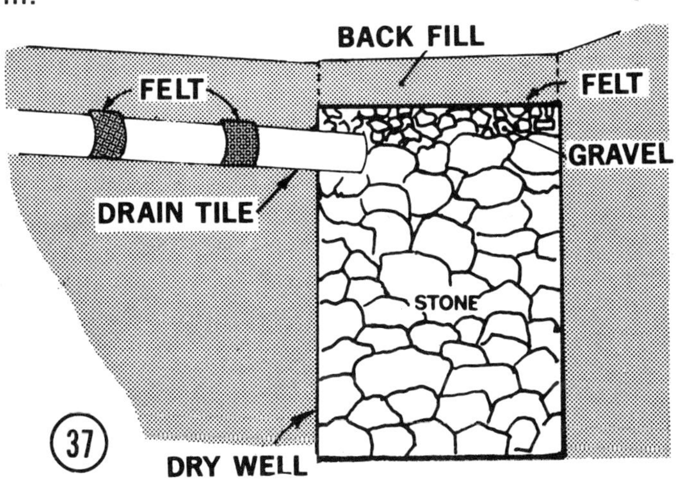

TERMITE SHIELD

Place a continuous length of 12" copper flashing, Illus. 38, on top of foundation as a termite shield. 92 lineal feet of 12" copper flashing is recommended, Illus. 38.

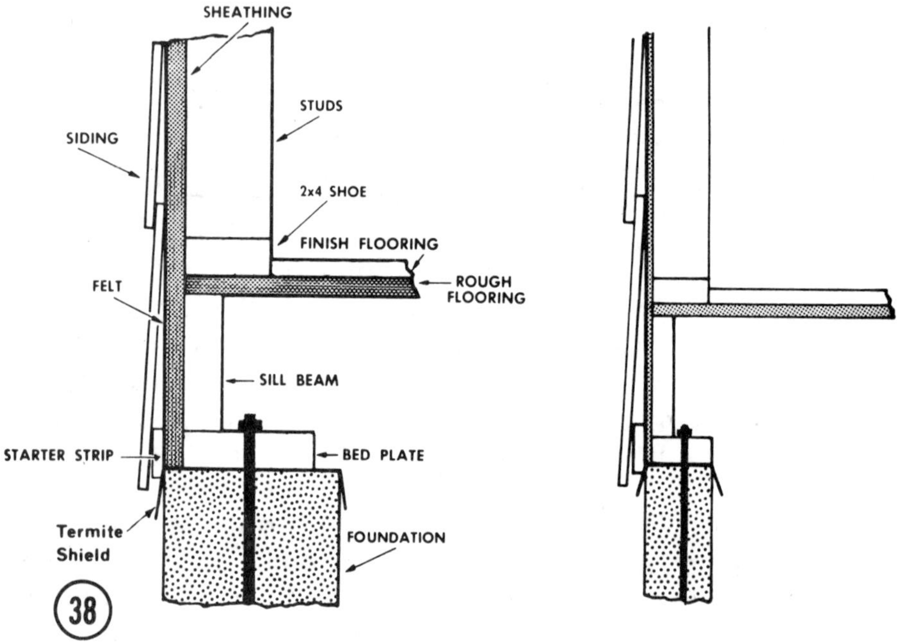

Cut flashing to overall length required and place it alongside anchor bolts. Using a ½" chisel, cut holes to receive anchor bolts. Make holes of sufficient size to be entirely free from contact with anchor bolts. Insulate bolt with asphalt. Run a continuous ribbon of flashing over window opening and around entire foundation wall. The termite shield should project 2" over each side of block wall. Bend termite shield as shown, Illus. 38. Seal each bolt hole with asphalt.

F.H.A. and many local codes now permit ⅜" exterior plywood siding without sub-sheathing. Your building material retailer can advise you.

PLEASE NOTE: While this book provides a simplified guide to building, due to the variance in lumber width and thickness, it does not assume responsibility for dimensions specified for inner framing members. Always measure, then cut framing to length your contruction requires.

LAY BEDPLATES

Use 2 x 8 — 2/10', 6/12'

When buying lumber, order as a "pro." Always specify size (2 x 8), then number of pieces followed by length. For example: 2 x 8 — 2/10 means two 10 ft.

To locate exact position of each anchor bolt, cut bedplate to length required, Illus. 39, Bedplate is recessed 3/8" from outside face of block all the way around.

Place bedplate in position alongside anchor bolts, Illus. 40. With a square, draw exact thickness of each bolt on bedplate.

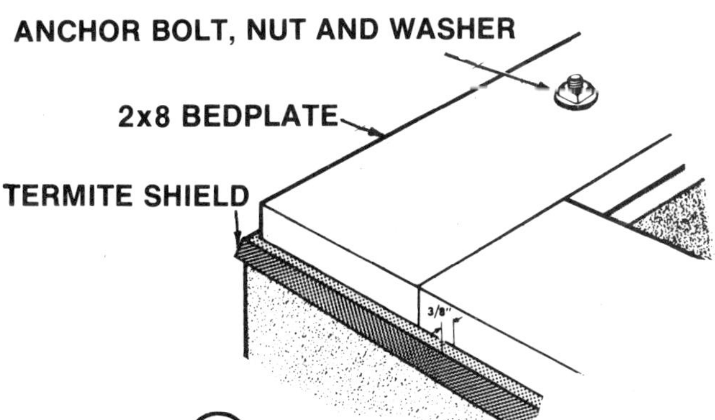

39

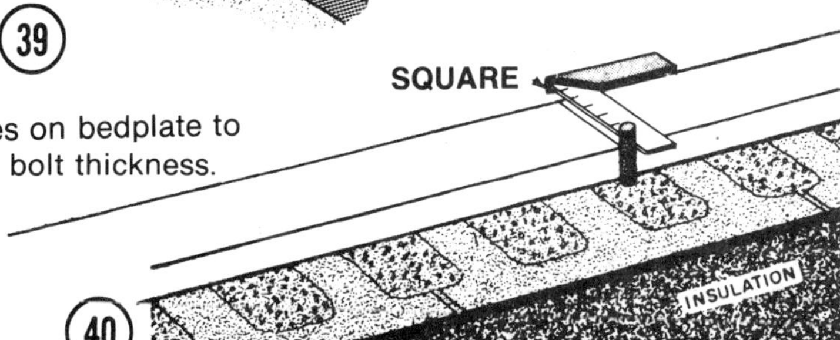

Draw lines on bedplate to represent bolt thickness.

40

To accurately position each bolt, use a scrap of 2 x 8 as a pattern. Since 2 x 8 measures 7¼", when placed flush with inside edge of block, it sets ⅜" from outside edge. Drill ¾" hole in pattern in exact position bolt requires. Then drill hole in bedplate. Holes can be drilled slightly larger if you find it necessary. Bedplates will be flush with inside edge of block, Illus. 39.

After drilling holes, place bedplates in position and check diagonals. Hold bedplates to outside dimensions of 20'7¼" x 24'7¼", or size desired. Toenail bedplates together at all butt joints.

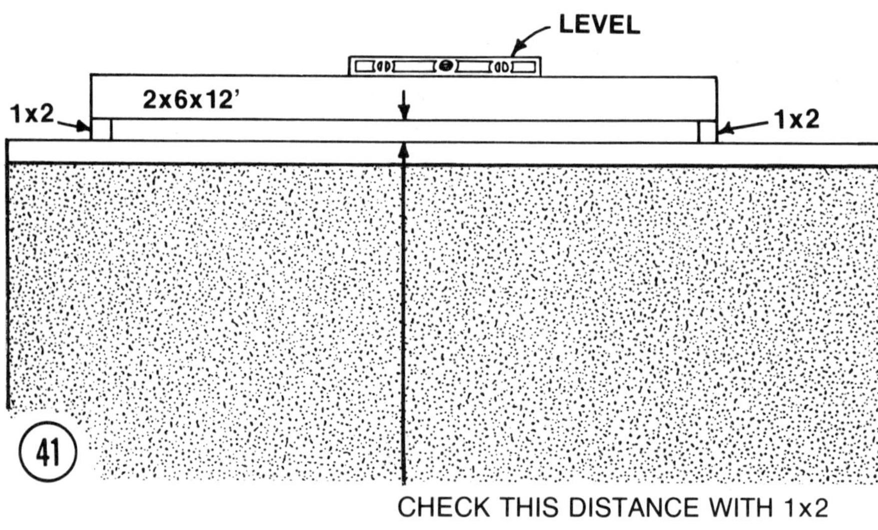

41

Level bedplates, Illus. 41. To simplify leveling, nail pieces of 1 x 2 to edge of a straight 2 x 6 x 12'. Place a level on top. Test space between bottom edge of 2 x 6 and bedplate with another piece of 1 x 2. Shim bedplate with a piece of slate or shingle where necessary to raise. When bedplate is level, place flat washers on bolts. Only finger tighten nuts at this time. Make a wet mixture of one part cement to two to three parts of sand and flush concrete under bedplate. Allow mortar to set three days, then tighten nuts.

FOOTINGS FOR LOAD BEARING BASEMENT PARTITIONS

Note Illus. 26. Dig trenches in position indicated. Set up 2 x 8 forms, Illus. 42. Footing for partition curb should be approximately 8'' wide, 7¼'' deep. 2 x 4 framing can be used in a basement partition. Use 4 x 8 x 16'' block for partitions in an earth sheltered house.

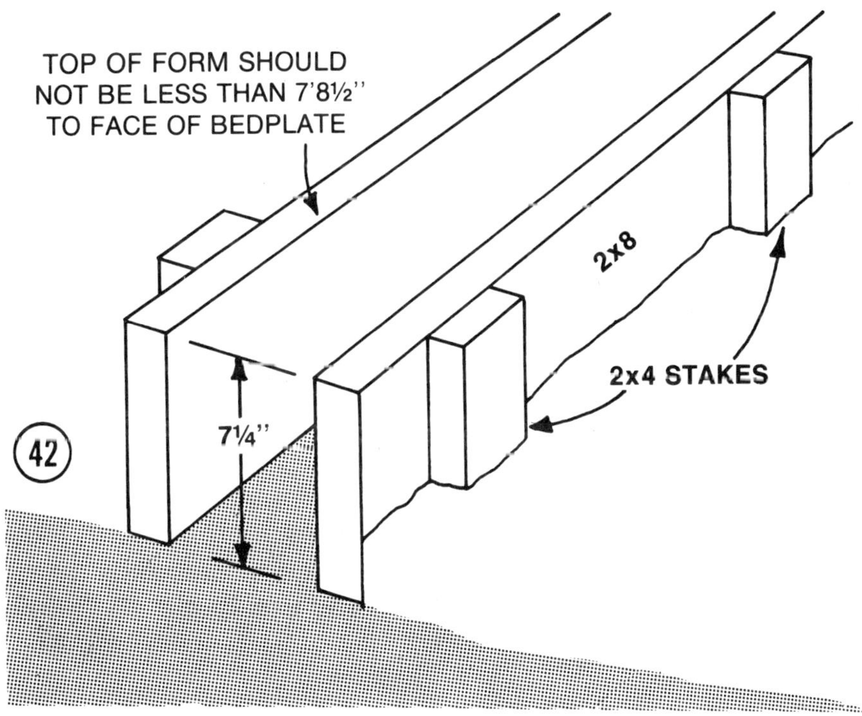

Top edge of form must be level. The exact height to top of footing form can be determined by measuring down foundation wall from top of bedplate. Stretch a line between foundation walls, Illus. 24.

Pour curb footings with 1 part cement, 3 parts sand to 4 to 5 parts of run-of-bank gravel. Pour footing on undisturbed soil. Allow footing to set at least three days before starting work.

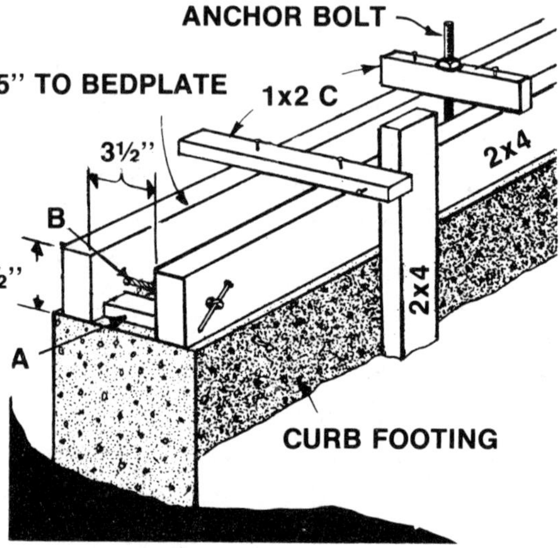

43

A 1x2 SPREADER
B TIE WIRE

Build 2 x 4 forms for partition curb, Illus. 43. Use pieces of 1 x 2 for spreaders A. Use tie wires B, or hold forms with 1 x 2 C. Remove 1 x 2 spreaders as poured concrete holds forms in place. No curb is required when you use 4 x 8 x 16" block for partitions.

Illus. 32 shows location of anchor bolts in partition curb. Place ½ x 8" machine bolts in location noted. Hold bolts upright as shown, Illus. 43. Allow curb to set at least three days before removing forms or building partitions. Always measure overall height required for interior partitions before cutting studs.

NOTE: When laying block for an interior partition, you can, if necessary, position a form on top course and pour concrete cap to height joists require.

PORCH FOUNDATION

Note Illus. 24,26,32,44. Set up batter boards and stretch guide lines for porch foundation. Dig trenches 2' wide. Footings should be 16" wide, 8" deep, or size required by local codes. Footings should go down below frost level. The concrete block wall for porch is on same level as foundation of house. Estimate how many 8" courses of block will be required, and lay footings to required height.

CEMENT 1" GALV. PIPE IN POSITION

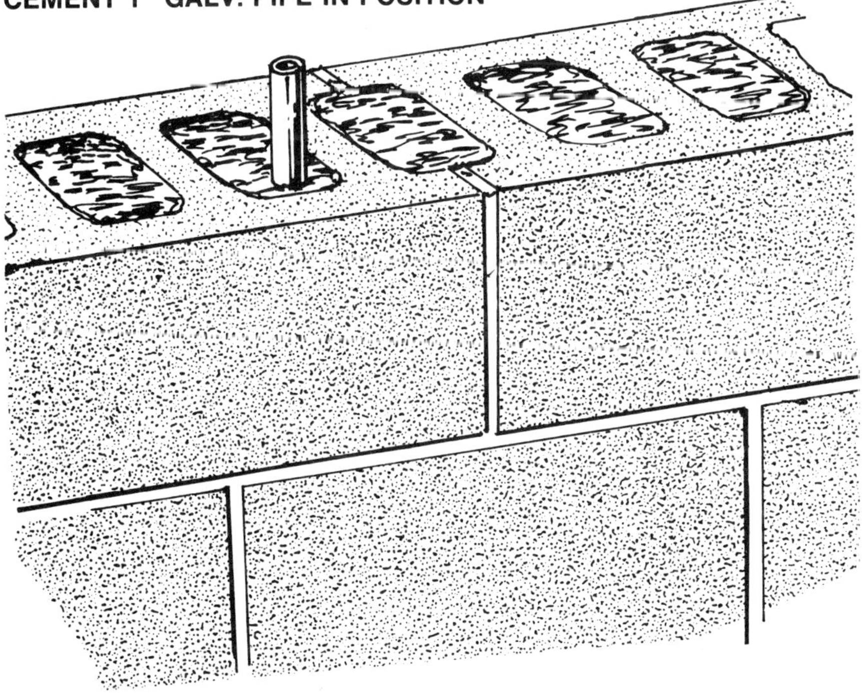

STUFF PAPER IN BOTTOM OF CORE TO HOLD CONCRETE

(44)

To anchor porch posts, place four 12" pieces of 1" galvanized pipe in position noted, Illus. 32,44. Allow approximately 4" of pipe to project above concrete block. Fill cores with concrete to hold pipe in position indicated, Illus. 32.

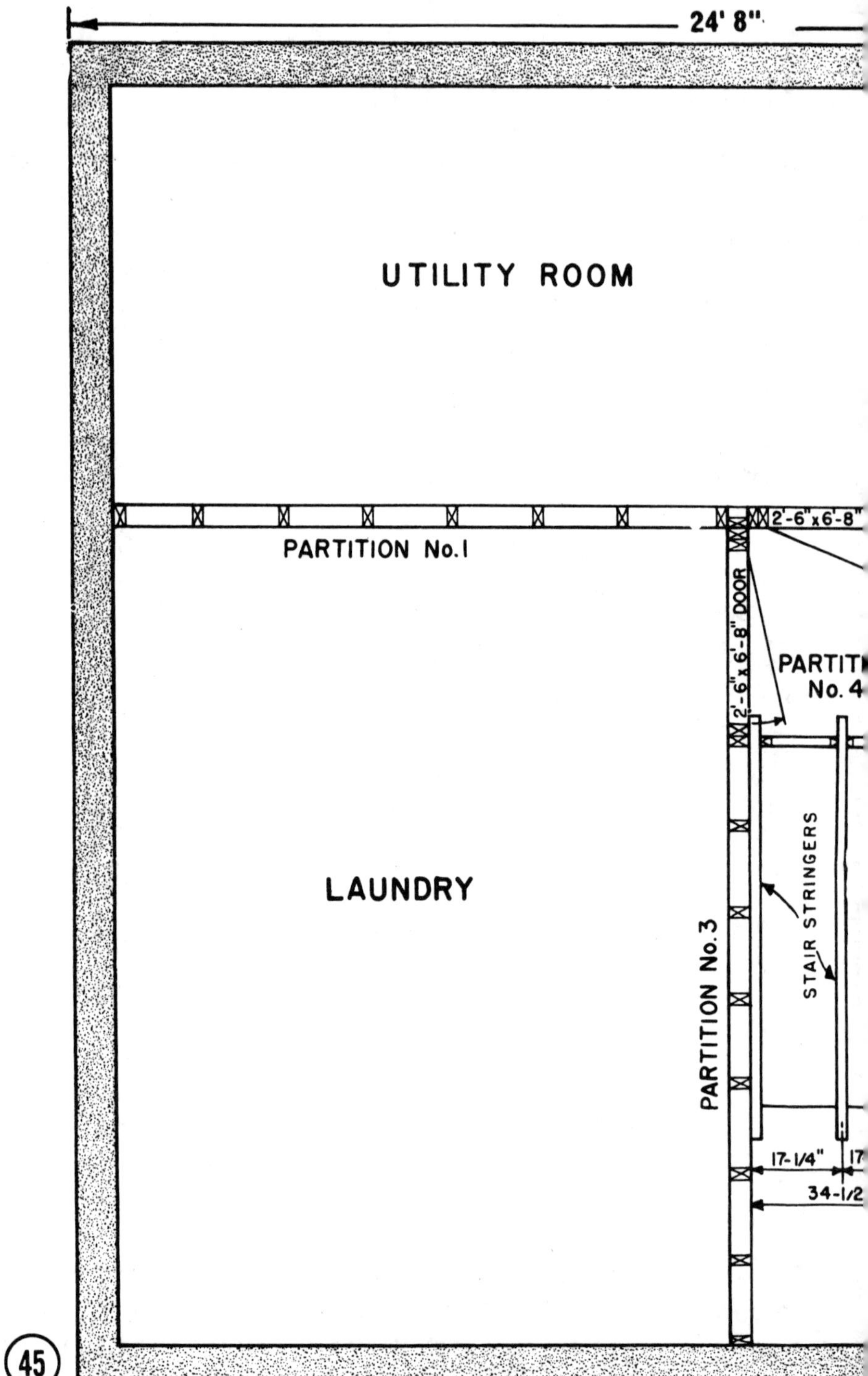
24' 8"
UTILITY ROOM
2'-6"x6'-8"
PARTITION No.1
2'-6"x6'-8" DOOR
PARTITI
No. 4
LAUNDRY
PARTITION No.3
STAIR STRINGERS
17-1/4"
34-1/2
45

PORCH

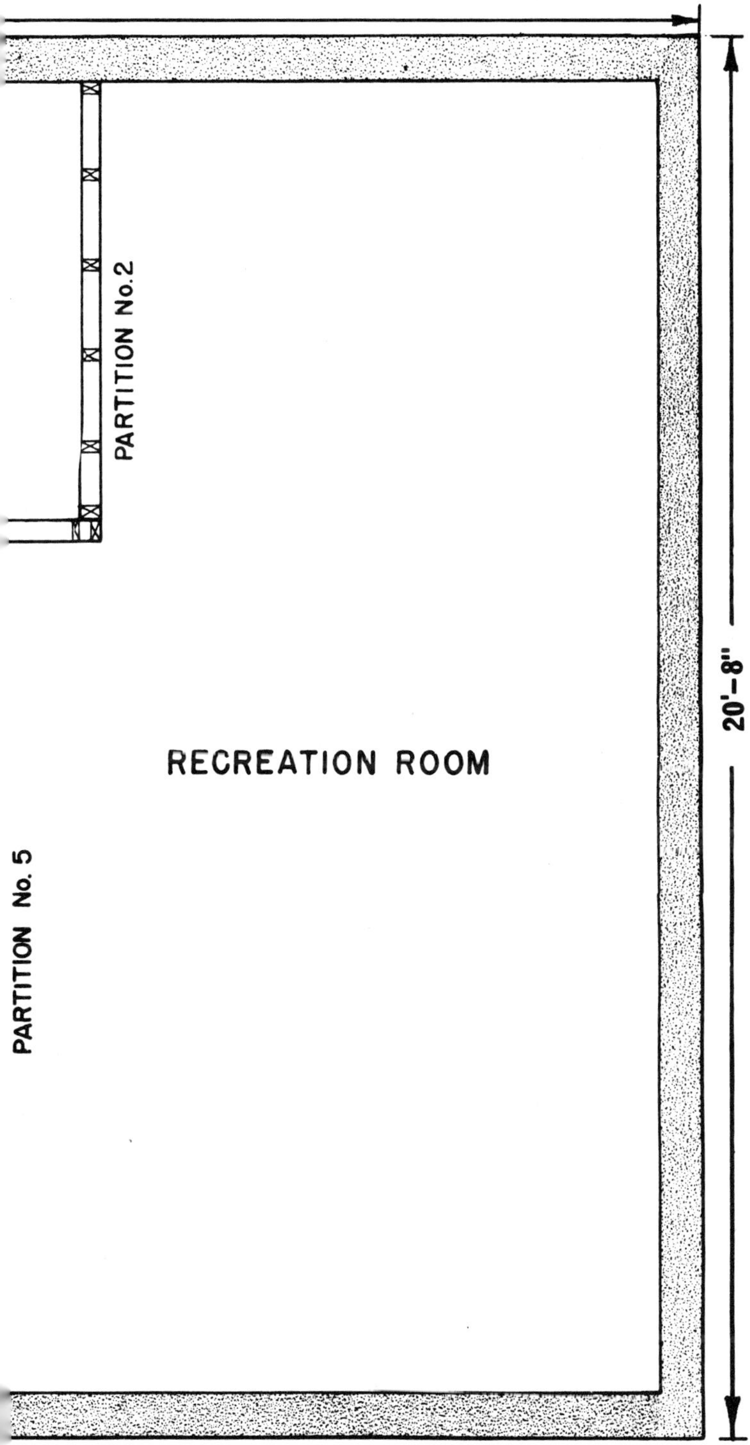

PARTITION No.2
RECREATION ROOM
PARTITION No. 5
20'-8"

BASEMENT PARTITIONS

Note Illus. 45,46,47,48,49,50,51,52. Partitions 1,2 and 3 are assembled on floor, then raised and braced in position. Always check end of lumber with a square, Illus. 52. Cut end square before measuring length required.

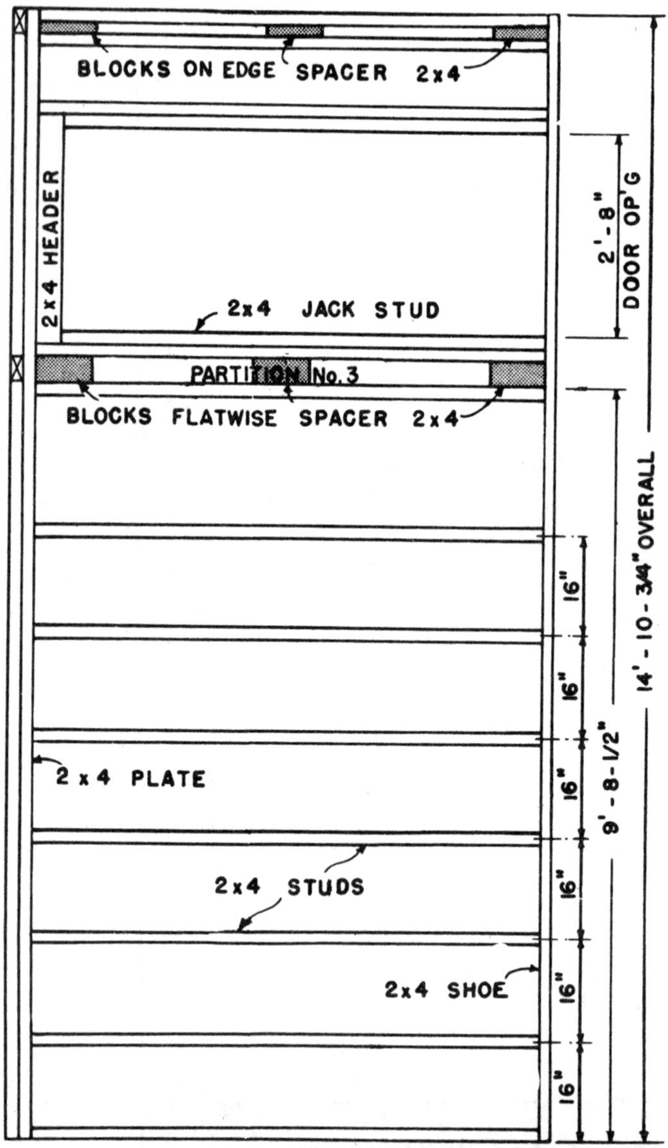

PARTITION No. I

BUILD ALL PARTITIONS TO OVERALL HEIGHT AND LENGTH REQUIRED

(46)

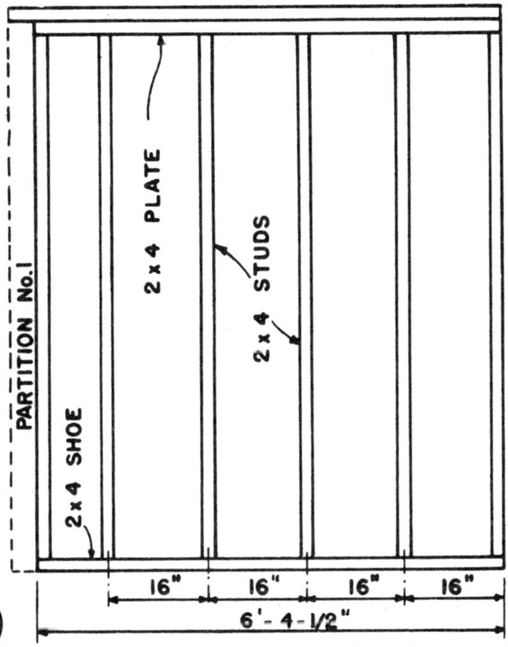

47

PARTITION No. 2

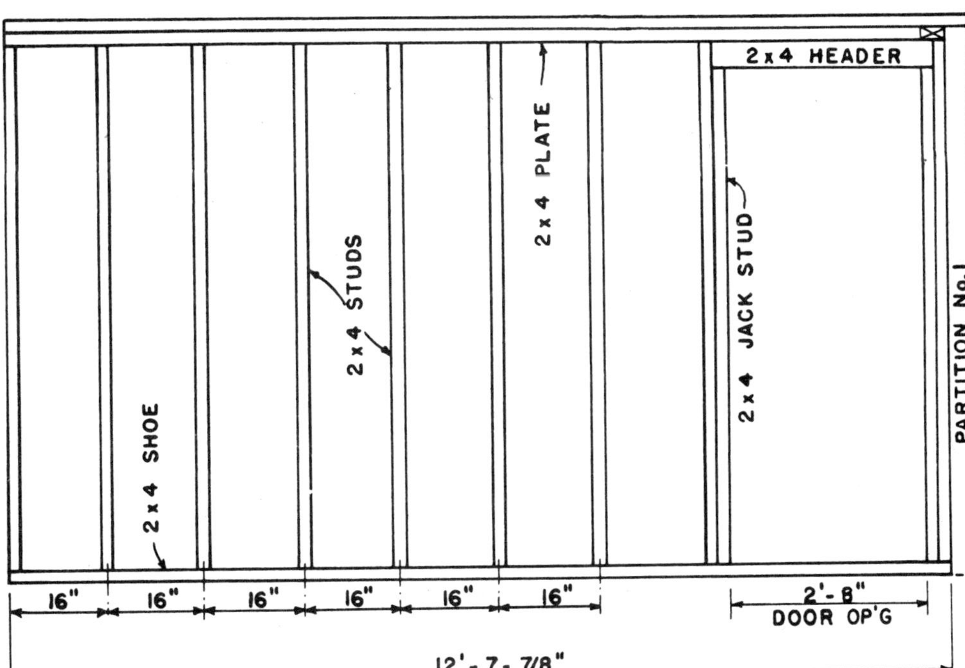

PARTITION No. 3

48

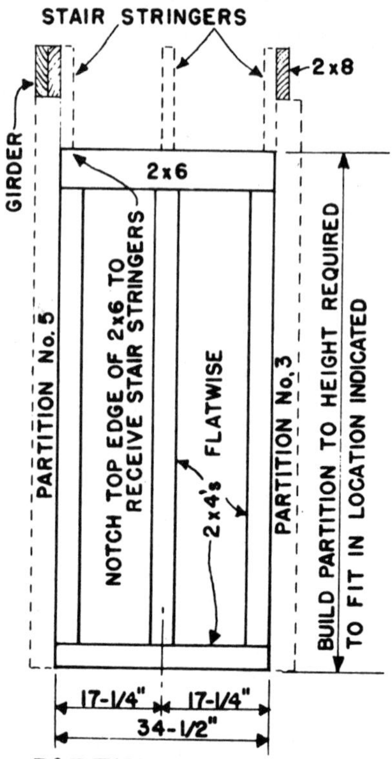

PARTITION No. 4

(49)

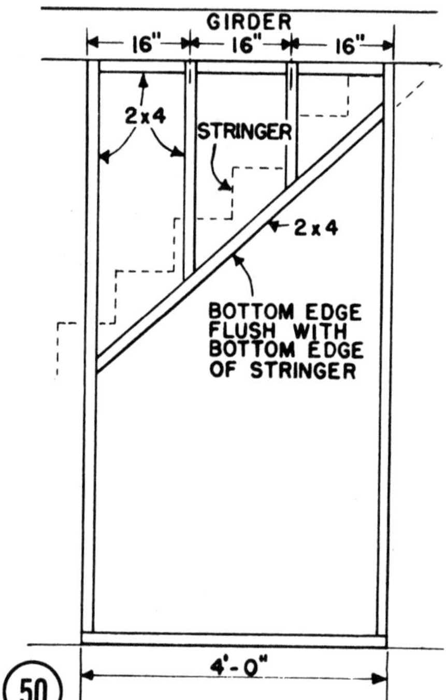

(50)

PARTITION No. 5

(51)

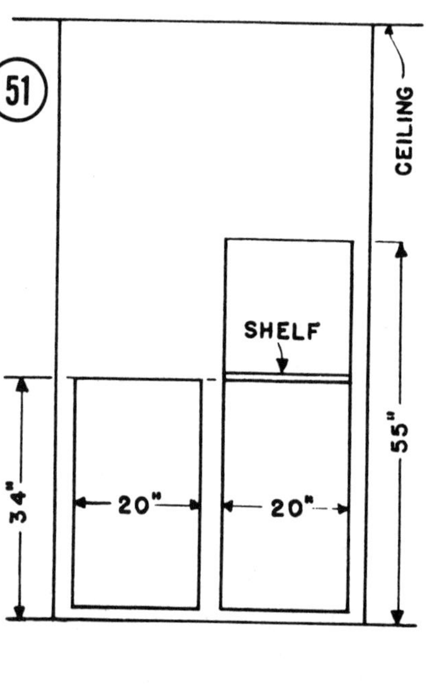

DIMENSIONS OF CLOSET UNDER STAIRS

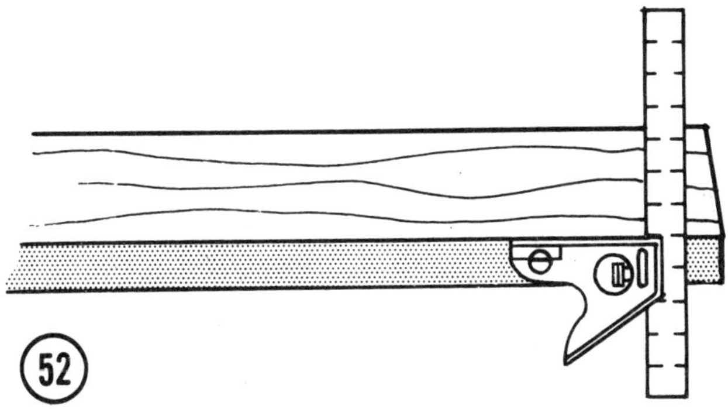

52

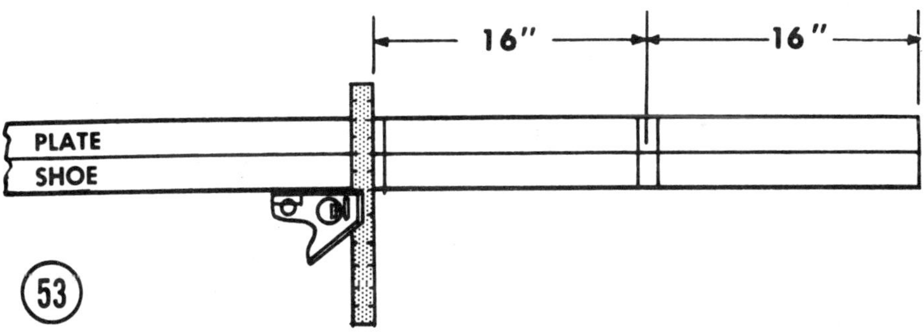

53

Cut shoe and plate to length required. Place them side by side, check ends with square, Illus. 52,53. Measure and draw position of each stud across shoe and plate. Nail through shoe and plate into each stud using two 16 penny common nails. Insert and nail spacer blocks where indicated. Use any scrap lengths of 2 x 4 approximately 8" or longer for spacer blocks. NOTE: Some spacer blocks are placed flatwise, some on edge as shown in each illustration.

After assembling a wall frame, check diagonals. When diagonals are equal length, frame is considered square. If frame isn't square, hit it with a hammer at corner to drive into position. When diagonals are equal, nail a 1 x 4 diagonally across studs to hold square.

Raise and brace each partition in position in sequence numbers indicate.

Partitions 4 and 5 are built after concrete has been laid for rough basement floor and rough stairs have been installed.

After partitions are raised and braced in position, a second 2 x 4 plate is nailed in position. Top plate overlaps at corners, Illus. 46,47,48.

Illus. 50 suggests size of closet that can be built under stairs. A niche can be built to accommodate a TV set.

To build partition No. 1, Illus. 46, use 2 x 4 — 3/16, 7/14, 1/6. Cut 2 x 4 shoe and plate 14'10¾". Place shoe in position along curb, mark location of anchor bolts following same procedure used for bedplate, Illus. 40. Cut 2 x 4 studs to length required so top 2 x 4 plate is level with bedplate on foundation.

Cut and nail double 2 x 4 header over door opening, Illus. 46. Cut 2 x 4 jack studs at door opening to length required and nail in position indicated.

Place assembled partition on curb. Place washers on anchor bolts and fasten nuts "finger tight."

To ascertain whether a partition is plumb, drive a nail into plate, Illus. 54, and tie a plumb bob line. Place line 1" from end of plate. Now measure distance point of bob is from shoe. When point of bob measures same distance from shoe as it does from end of plate, end is considered plumb. Always plumb a wall at end and side. Brace partition in plumb position. Fasten to foundation walls with masonry nails driven into mortar joints or with wood shims under shoe. Toenail plate to joists, Illus. 57.

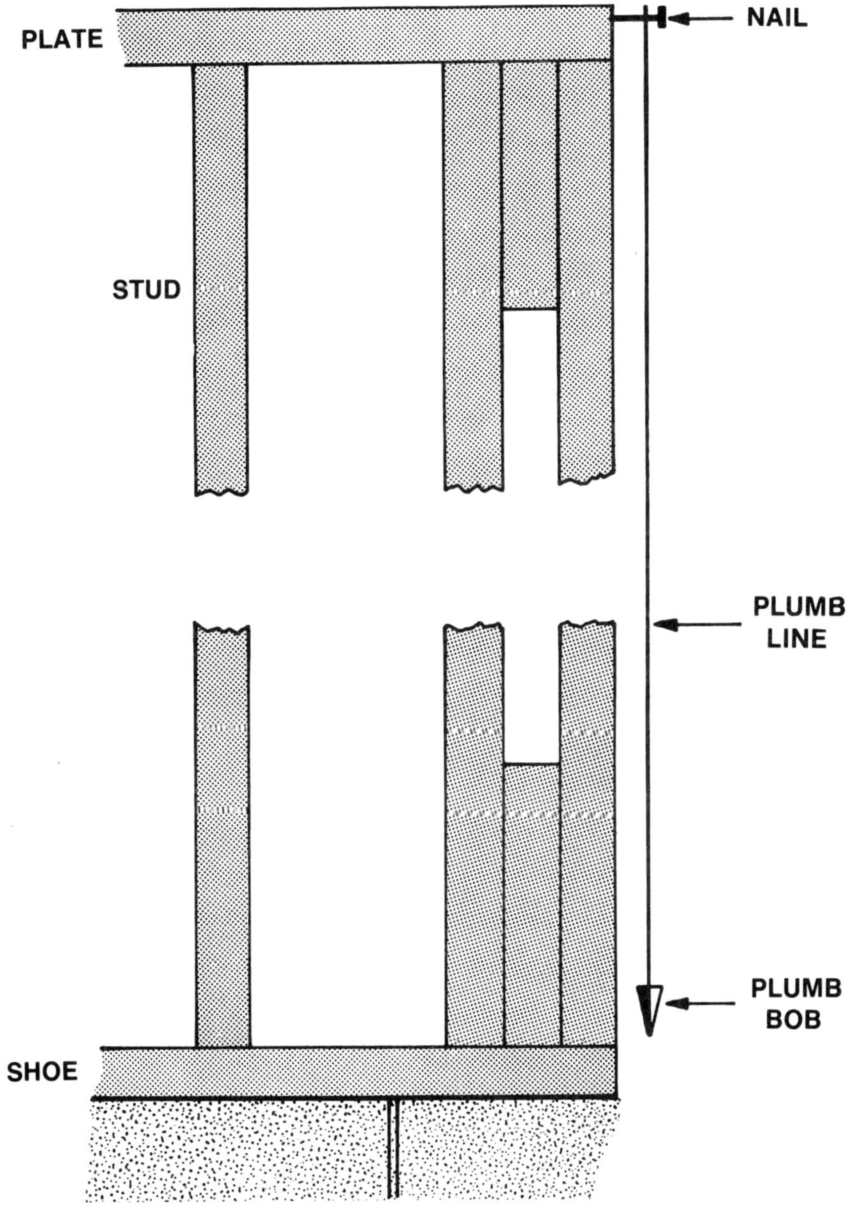

54

BUILD PARTITIONS 2 and 3

Use 2 x 4 — 1/20, 3/14 for Partition 2.
2 x 4 — 9/14 for Partition 3.

Note Illus. 47,48. Assemble, raise, plumb and brace partitions 2 and 3 following same procedure. Spike partitions together at corners.

To make certain partitions are straight, drive a nail into ends of plate and stretch a line, Illus. 55, from one end to other. Raise line at ends with 1 x 2. Use another piece of 1 x 2 to test space between plate and line at various points. Brace partition straight.

Cut top 2 x 4 plate to length required and nail in position. Top plate on partition 2 and 3 overlaps partition 1. Nail through top plate into lower plate.

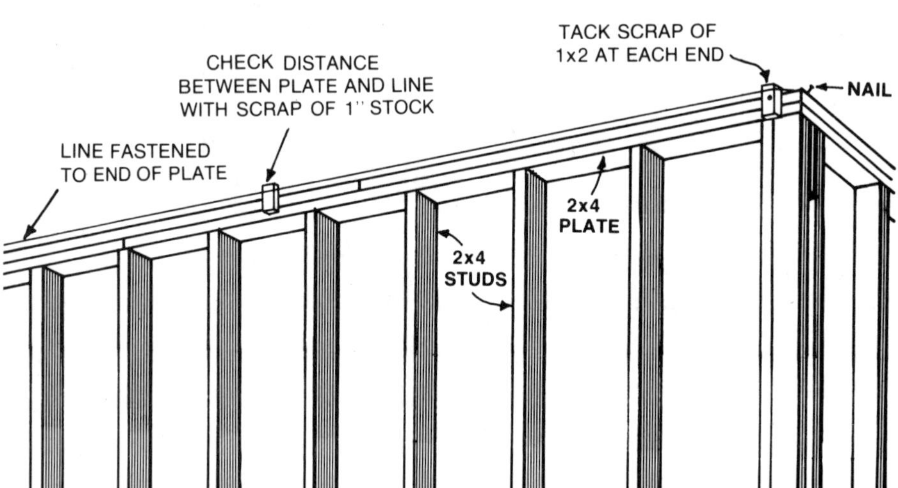

55

NAIL SILL BEAMS

Use 2 x 8 — 2/14,4/12,2/10.

Note Illus. 56,57,59. Cut 2 x 8 to length required and nail in position shown flush with edge of bedplate. Drive two 16 penny nails into each butt joint. Toenail sill beam to bedplate with 8 penny nails. Check corners with a square.

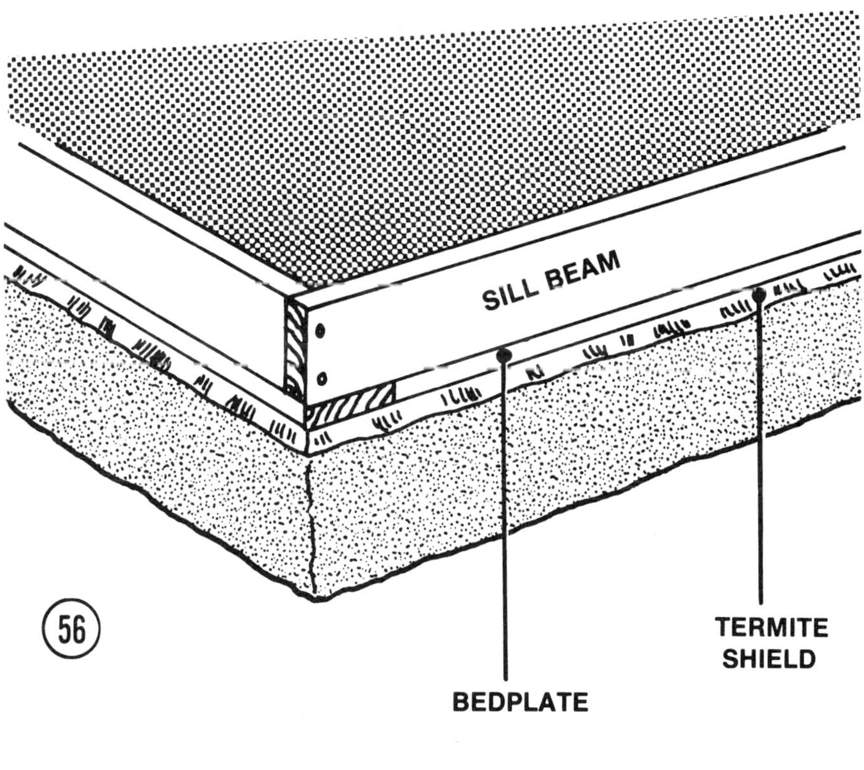

56

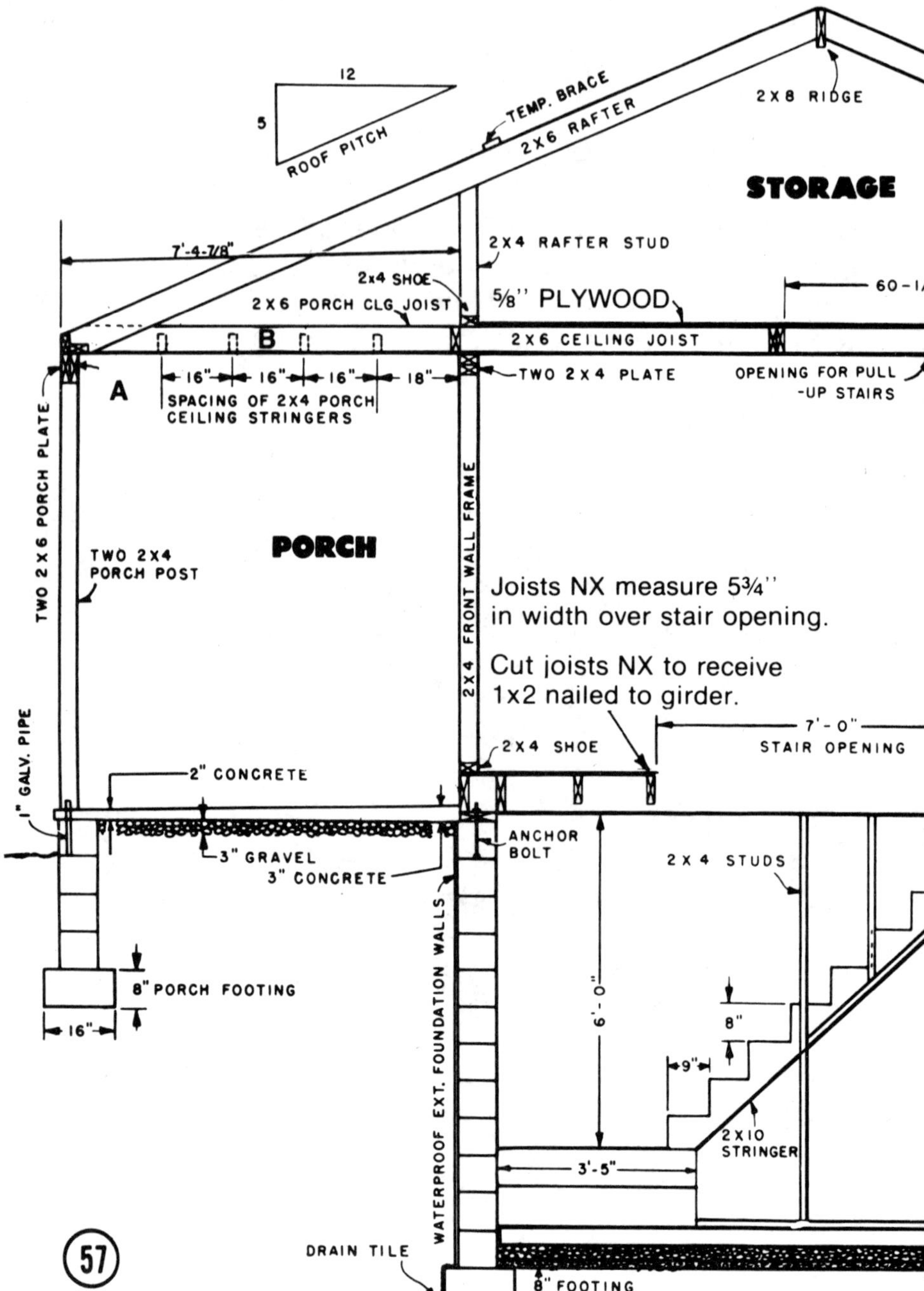

57

If an extra course of block has been added to foundation, add an extra 8'' riser and step to length of stair stringers. This should provide 6'8'' at base of stairs.

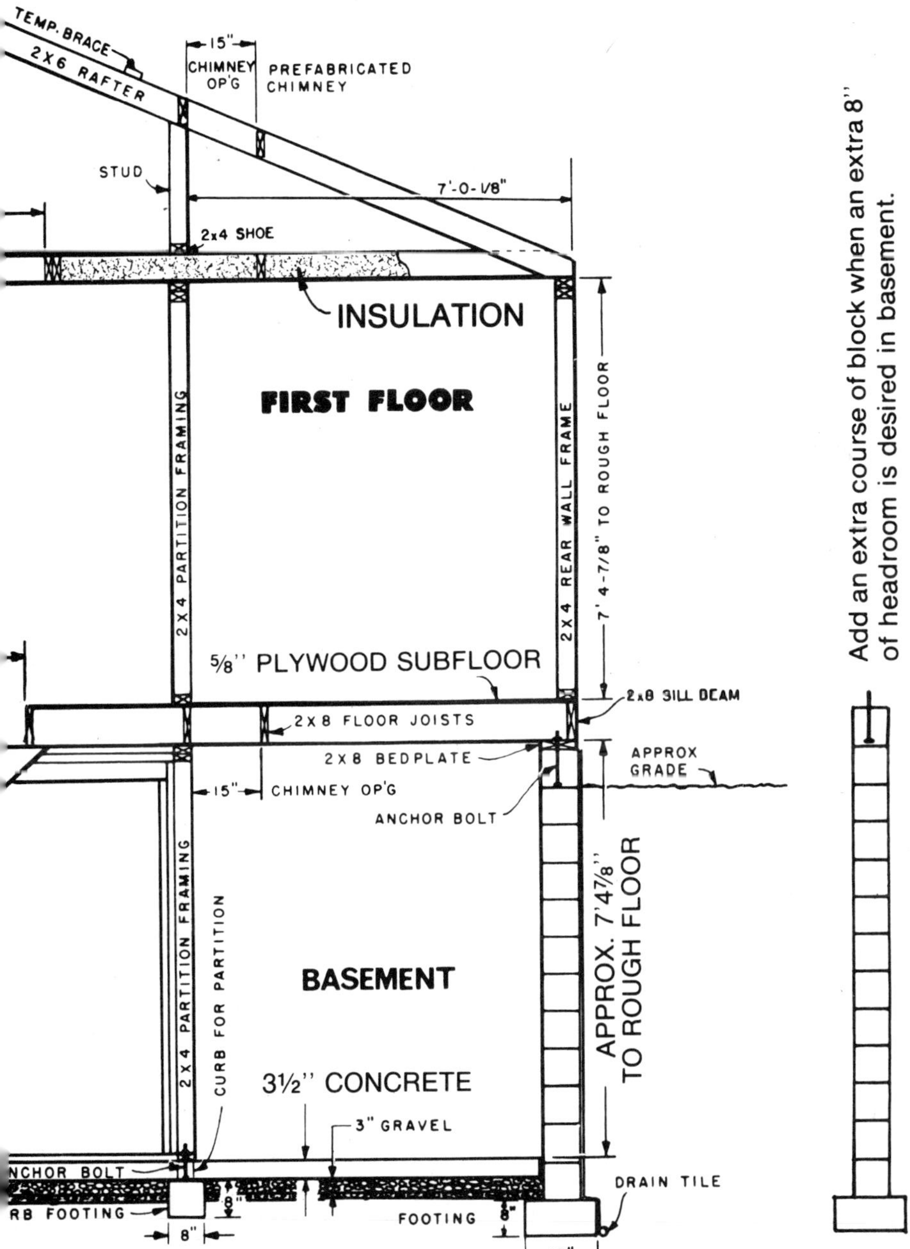

Concrete floor can also be laid on shoulder of footing.

SECTION VIEW OF EXPANDED HOUSE

58

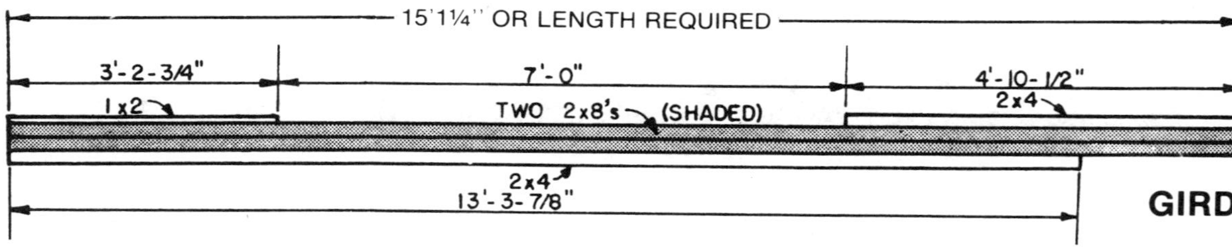

GIRDER

Cut joists NX to receive 1 x 2 nailed flush with bottom of girder.

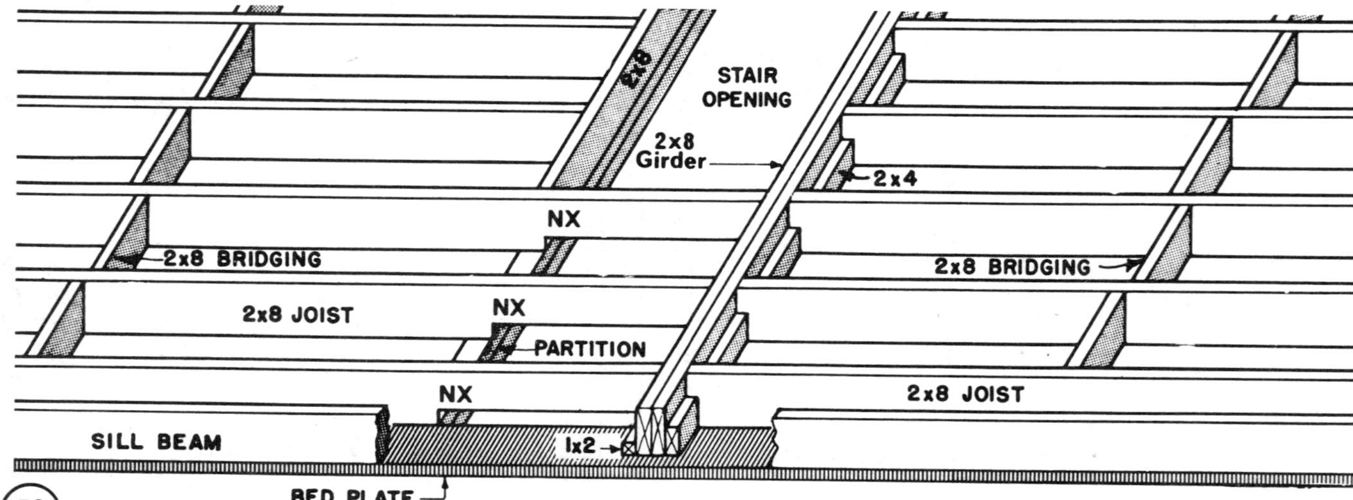

59

Sill beam is cut away to show position of 1x2 and 2x4.

GIRDER

If you alter size of house and need a longer girder, or if a partition is moved, or if you want entire area open, extend girder full length of house and support with steel lally columns or adjustable jack posts, Illus. 8,154.

Note Illus. 58,59,60. Nail two 2 x 8 x 15'1¼" together for girder, Illus. 58. Nail a 2 x 4 x 13'3⅞" to right side, a 2 x 4 x 4'10½" to left side and a 1 x 2 x 3'2¾" to left side where indicated. Nail sill beam to girder in position shown, Illus. 59,60. Toenail girder to partition.

FLOOR JOISTS

Use	2 x 8 — 6/16,6/14,15/12,11/10.	(Actual size		1½ x 7¼")
	2 x 4 — 1/20.	("	"	1½ x 3½")
	1 x 2 — 1/4.	("	"	¾ x 1½")

Note Illus. 59,60. The number alongside each joist, Illus. 60, indicates the length of lumber required.

With square, draw location of floor joists on girder, also on sill beam and on plate of partition, Illus. 53.

Always sight down floor joists. If curved, keep curved edge up. Notch floor joists to receive 1 x 2 or 2 x 4 nailed to girder, Illus. 59.

Notch joists NX, Illus. 59,60, to 5⅝" or width required where they project over stair platform. This provides additional headroom over stairs.

Cut joists to lengths required and nail in position indicated. Nail sill beam to joists with 16 penny nails at each joint. Toenail joists to bedplate, girder and to partition with 8 penny nails. Spike joists together where the 16 and 10's butt over partitions. Use short pieces of 1 x 6 or ⅝" plywood as gusset plates on both sides to reinforce butt joint.

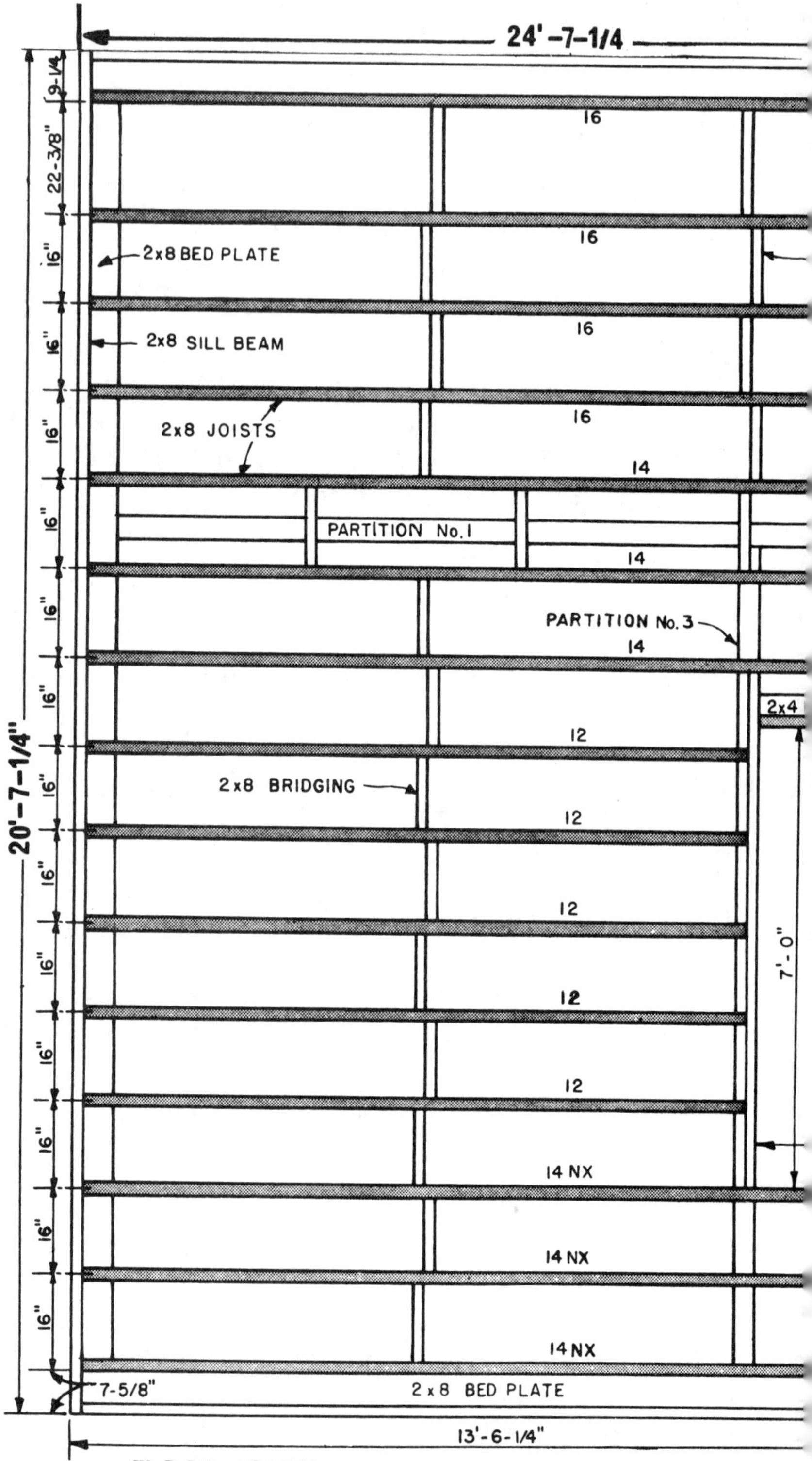

60

FLOOR JOISTS

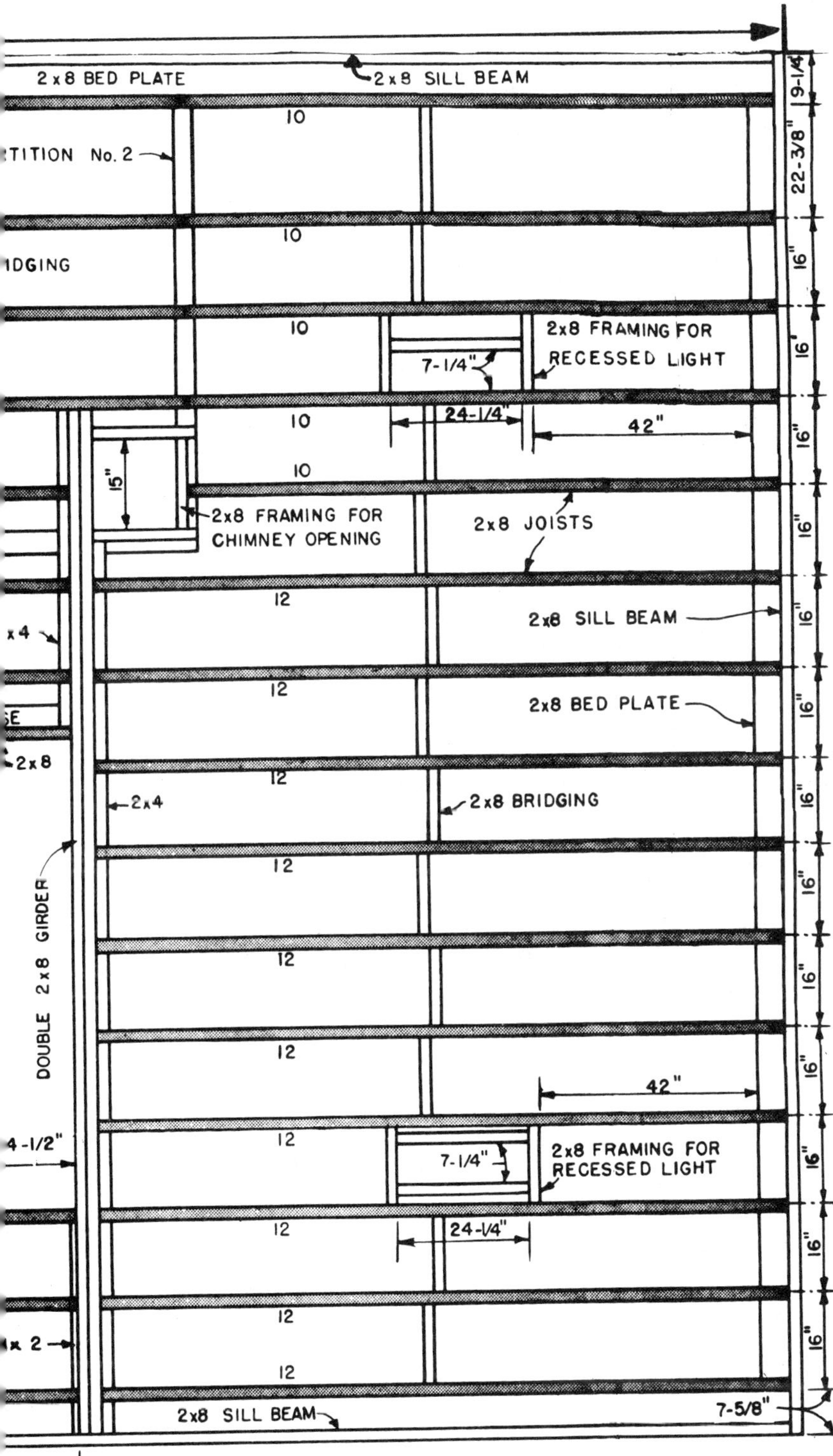

NOTE: ALL JOISTS ARE SHADED

PORCH

Nail 2 x 8 bridging in position shown, Illus. 60. Nail joists to bridging with 16 penny nails. You can use solid bridging, steel or spruce cross bridging, Illus. 61.

Frame openings for recessed lights if same are desired, Illus. 60; also for a prefabricated chimney. Follow manufacturer's recommendations.* Nail 2 x 4, 2 x 6 or 2 x 8 cats between joists where light fixture or partitions require a support.

SUBFLOORING

Note Illus. 56,57. Use ⅝ or ¾'' plywood to cover 525 sq.ft. 17 — 4 x 8' panels will be required. If you plan on covering subflooring with carpeting, use ¾'' plywood. Nail plywood with face grain across joists using 8 penny nails. Butt panels together over center of joists. Nail plywood subfloor flush with outside face of sill beam, Illus. 56.

If subflooring is to be left exposed for any length of time, cover with polyethylene.

*Also note installation described in Book #674 How to Install a Fireplace.

If no cellar is planned, those building in a cold climate should apply insulation between floor joists. Use thickness local climate requires. Mineral wool batts or blankets can be stapled between joists. This can be kept in place with hardware cloth (chicken wire) or wood lath stapled across bottom of floor joists. This insulation is important. Its cost is small compared to the comfort and fuel savings it provides.

FIRST FLOOR WALL FRAMES

Use 2 x 4 — 4/16,47/14,10/12,13/10.

Note Illus. 62,63,64,65,66,67,68. Illus. 62 shows position of frames. Assemble and erect each in numerical order indicated.

Nail a 1 x 4 wind brace on inside, Illus. 63, diagonally across from shoe to plate of each frame. This reinforces and holds frame square prior to application of sheathing. Keep temporary bracing on frame until sheathing has been nailed in place.

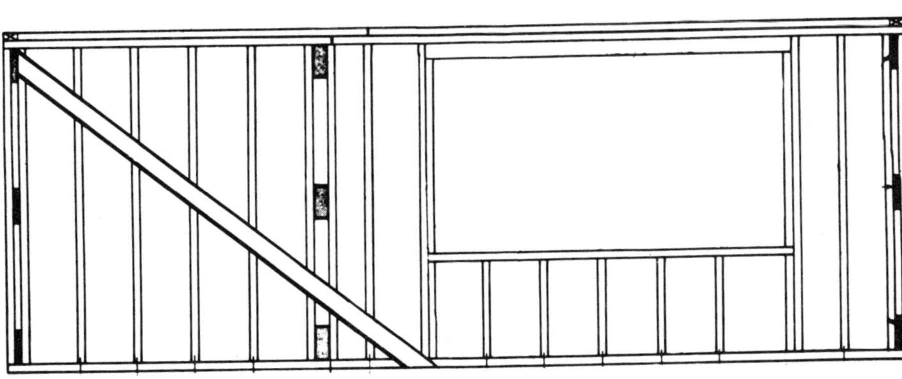

62

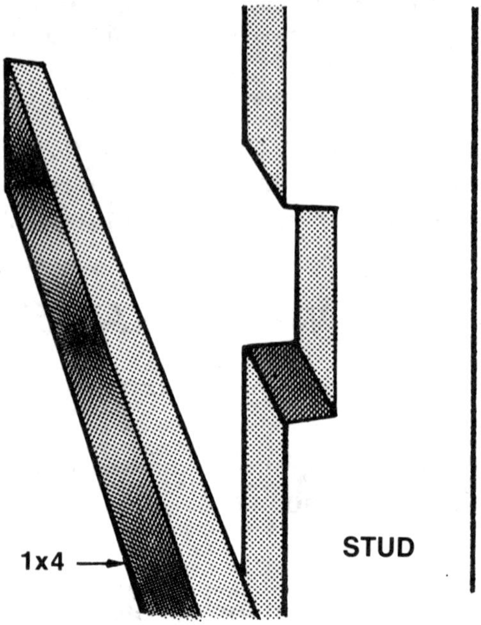

63

Next cut bottom end of 1 x 4 to angle required to fit against shoe. Place 1 x 4 in position diagonally across studs and cut top end to angle required to fit against lower plate.

Draw exact position across each stud. Remove brace and saw each stud to angle and depth brace requires, Illus. 63. Chisel out notch. Nail wind brace to each stud, shoe and plate with two 8 penny nails.

Build and raise right wall frame, Illus. 64, in position shown, Illus. 65. Nail assembled frame in place with 16 penny nails through shoe. Brace frame in plumb position. Build and raise left wall frame, Illus. 66; front, Illus. 67; and rear wall frame, Illus. 68. Always check each frame with a plumb bob in two directions. Nail top plate to assembled frames. Top plate overlaps at corners as indicated.

As shown in Illus. 55, check each frame to make certain it is straight.

Build partitions to height your construction requires.

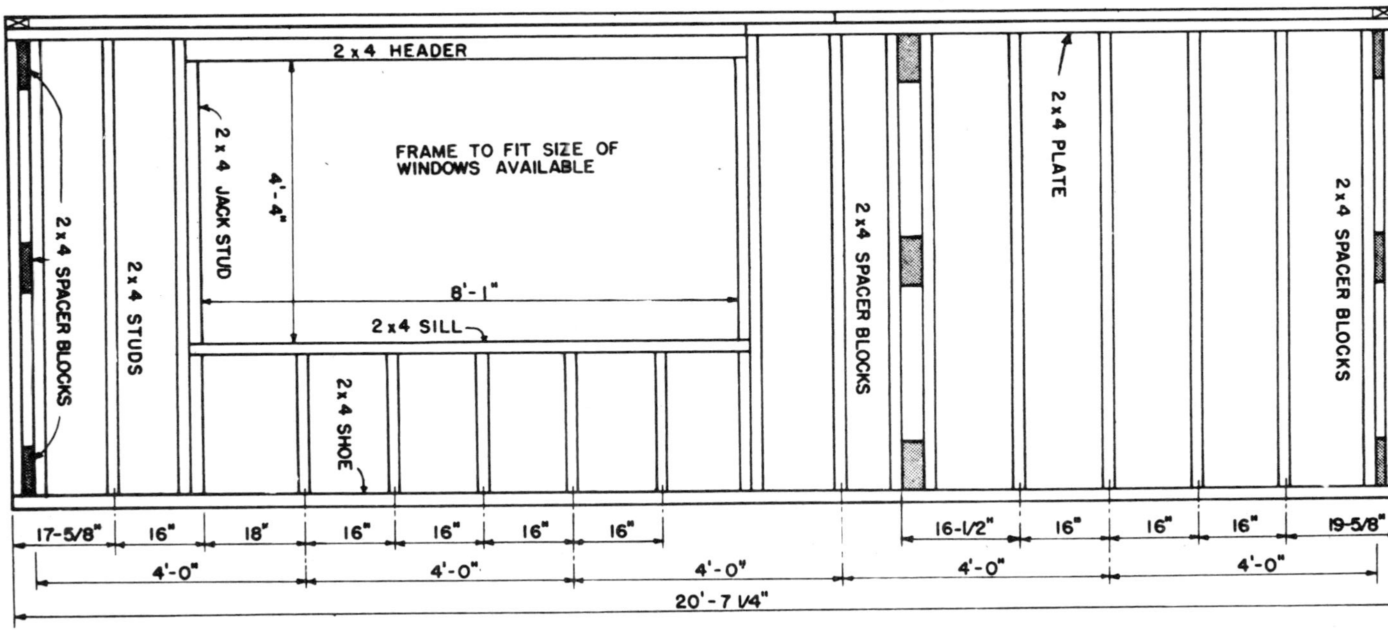

RIGHT WALL FRAME

64

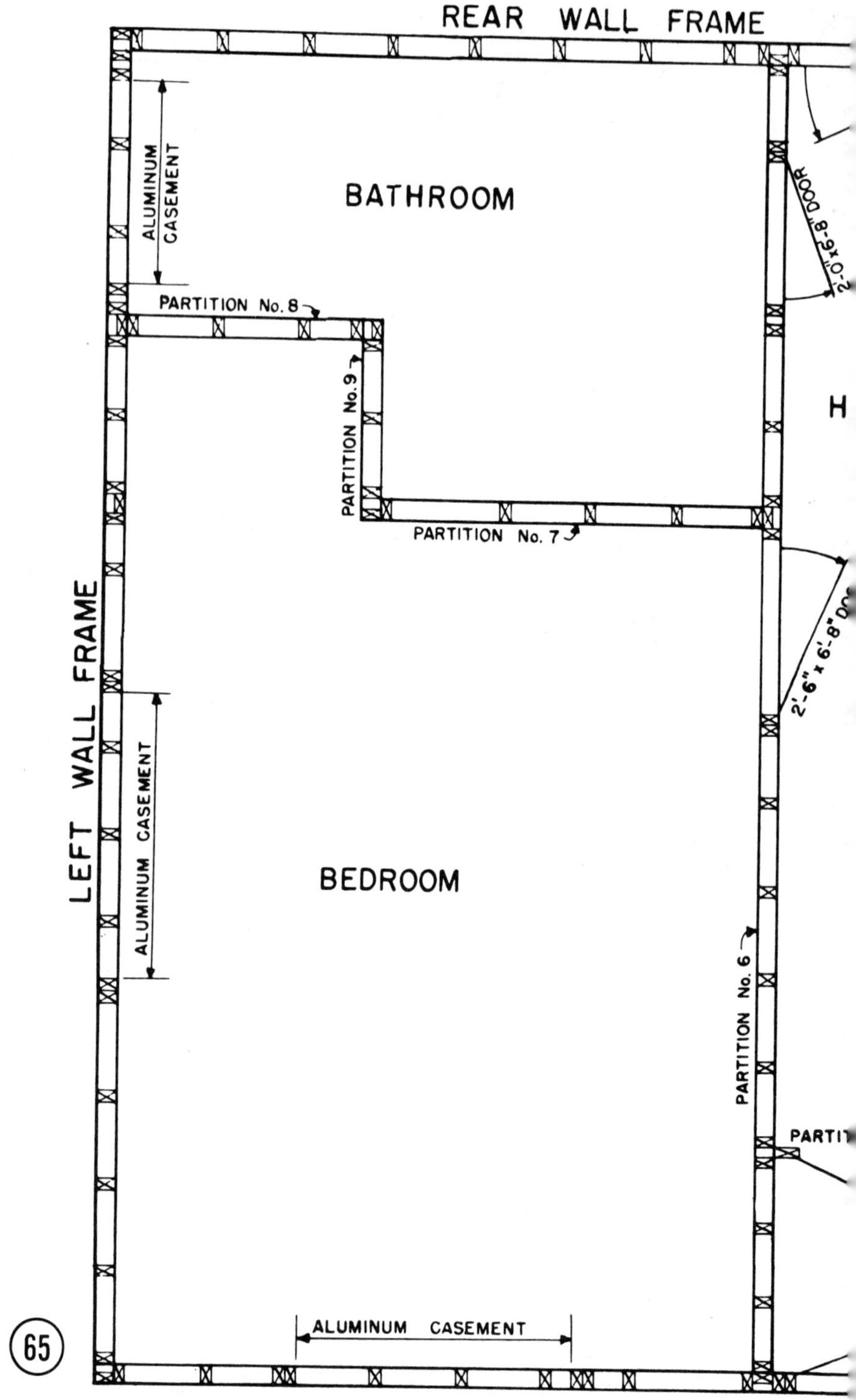
REAR WALL FRAME
ALUMINUM CASEMENT
BATHROOM
2'-0"x6'-8" DOOR
PARTITION No. 8
PARTITION No. 9
H
PARTITION No. 7
LEFT WALL FRAME
2'-6" x 6'-8" DO
ALUMINUM CASEMENT
BEDROOM
PARTITION No. 6
PARTIT
ALUMINUM CASEMENT
65

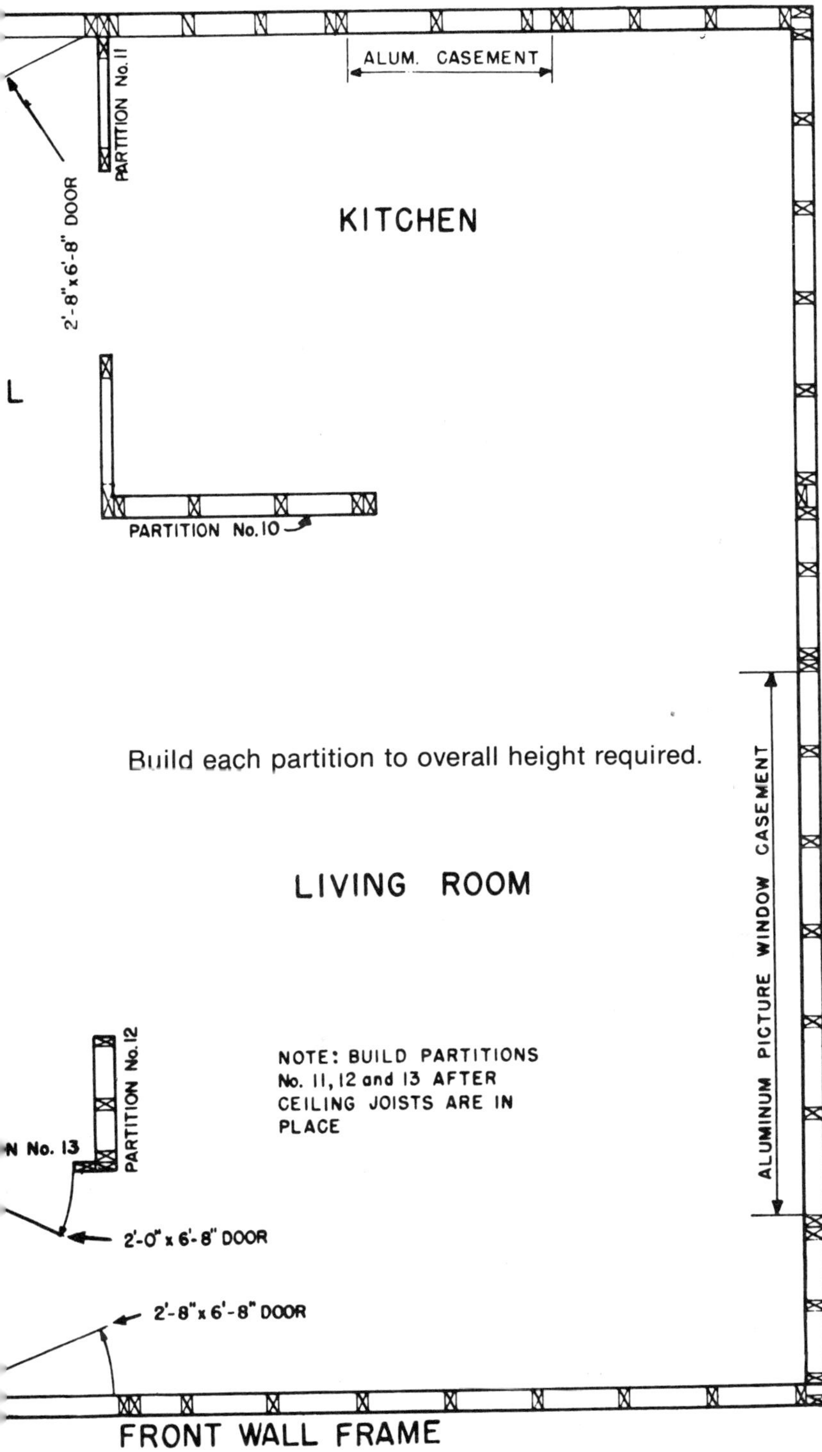
ALUM. CASEMENT
PARTITION No. 11
2'-8" x 6'-8" DOOR
KITCHEN
L
PARTITION No. 10
RIGHT WALL FRAME
Build each partition to overall height required.
ALUMINUM PICTURE WINDOW CASEMENT
LIVING ROOM
PARTITION No. 12
NOTE: BUILD PARTITIONS No. 11, 12 and 13 AFTER CEILING JOISTS ARE IN PLACE
N No. 13
2'-0" x 6'-8" DOOR
2'-8" x 6'-8" DOOR
FRONT WALL FRAME

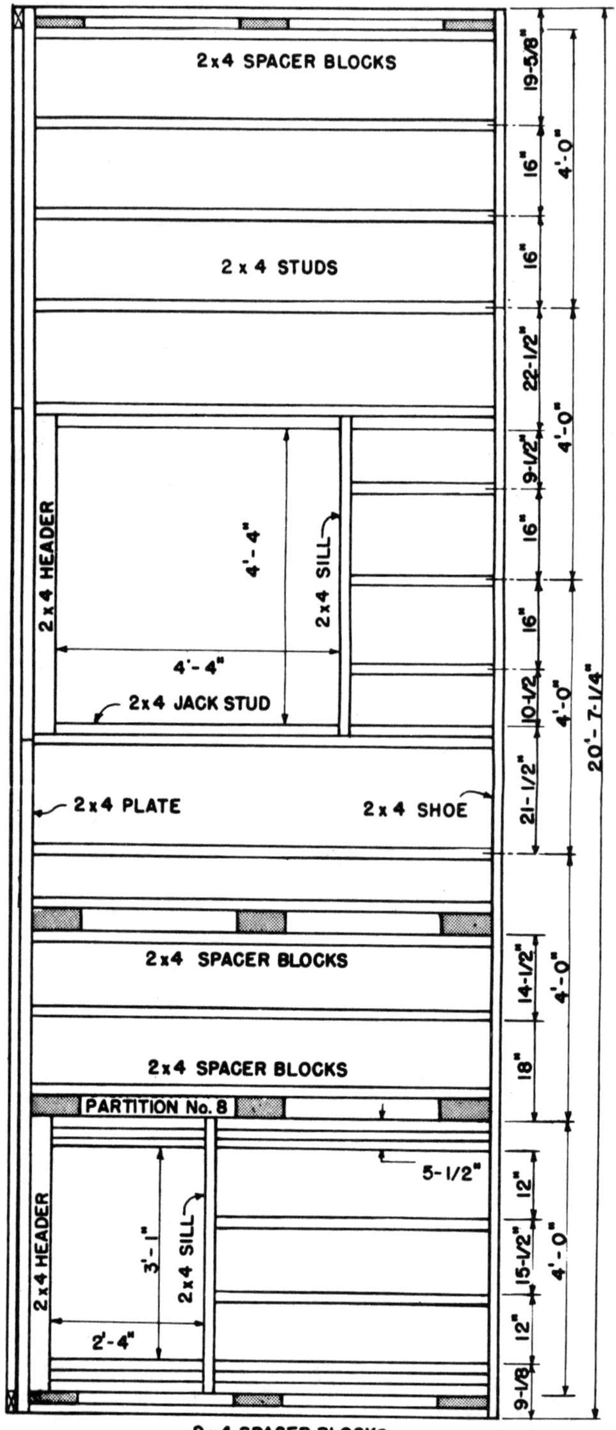

LEFT WALL FRAME

66

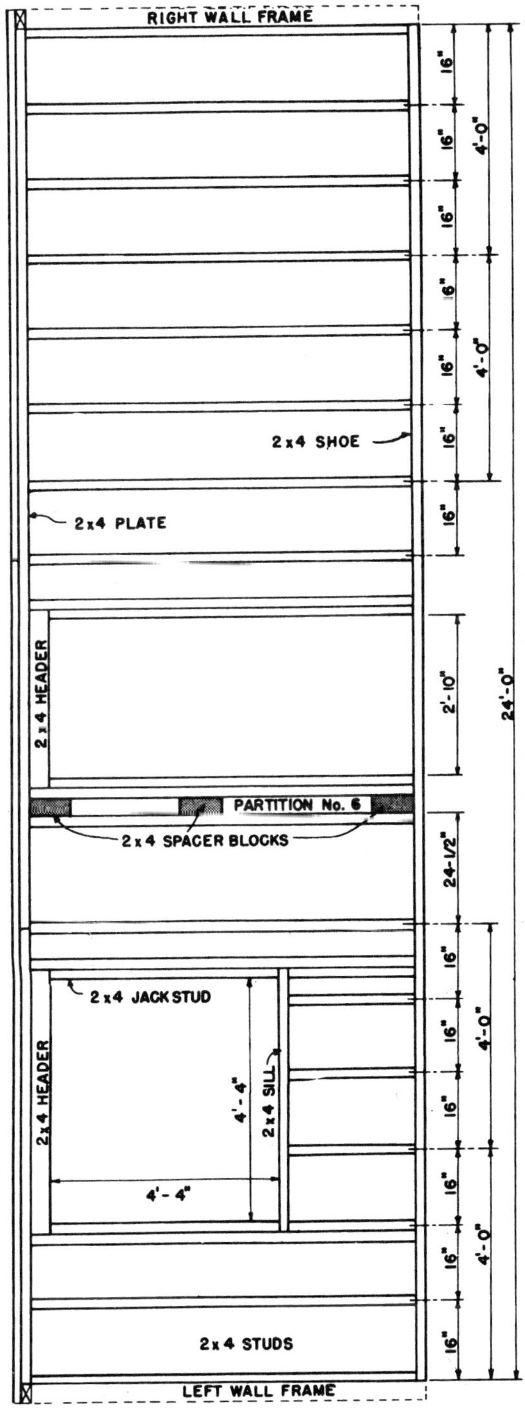

FRONT WALL FRAME

(67)

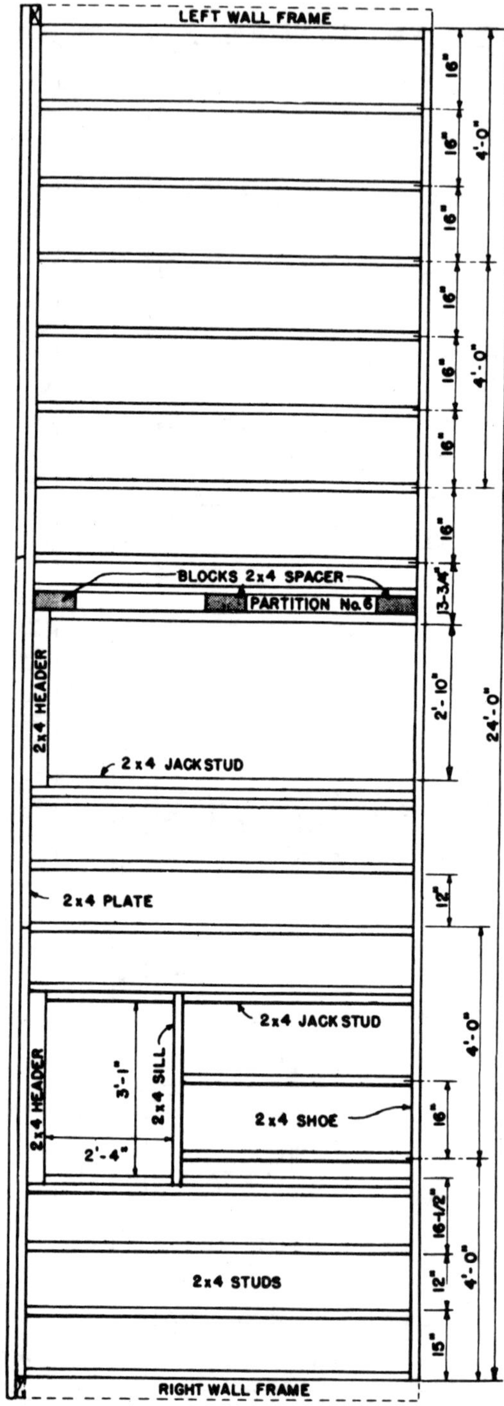

REAR WALL FRAME

68

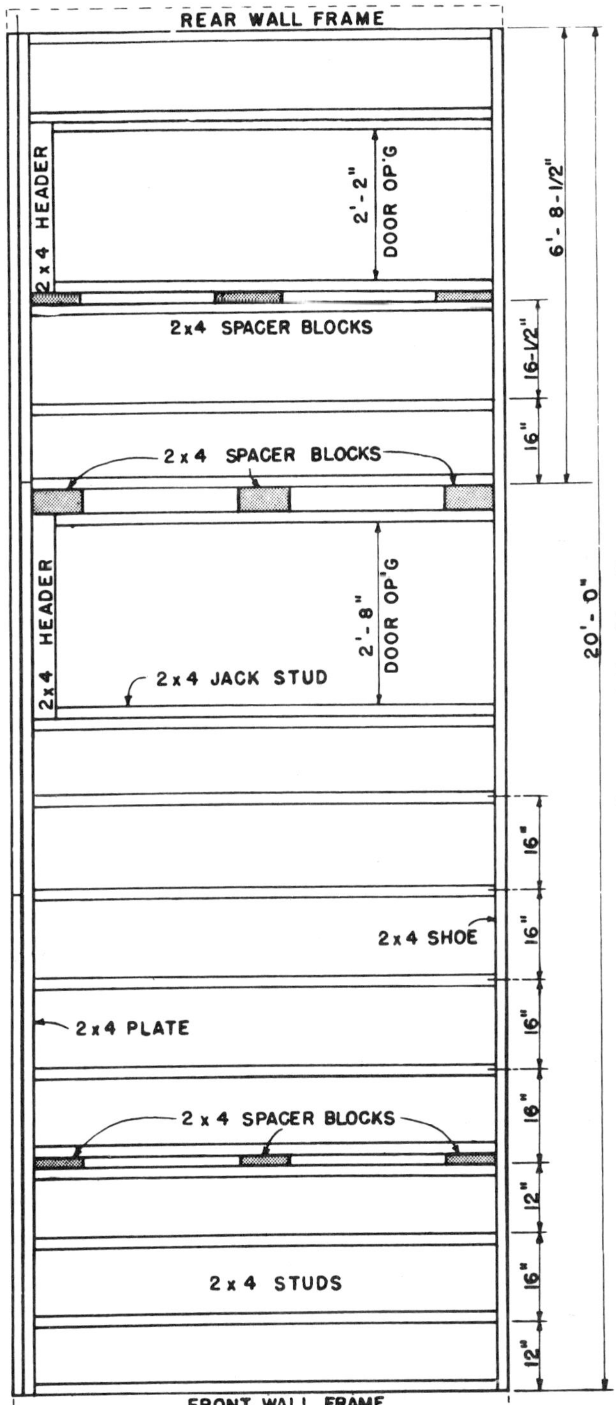

PARTITION No. 6

69

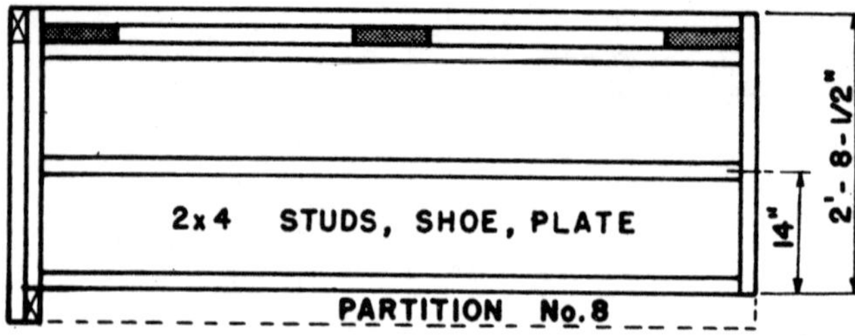

PART. No. 9

72

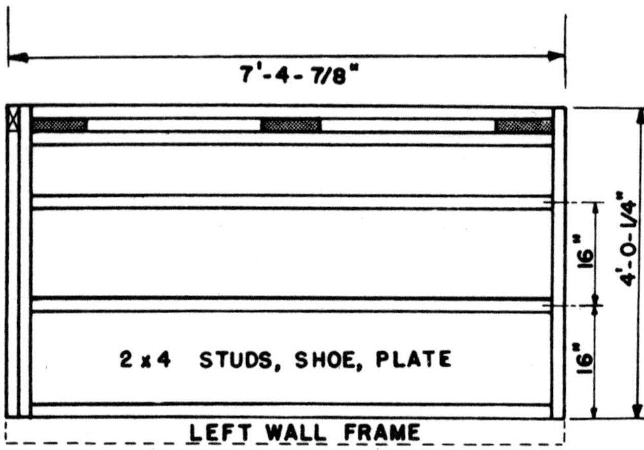

PARTITION No. 8

71

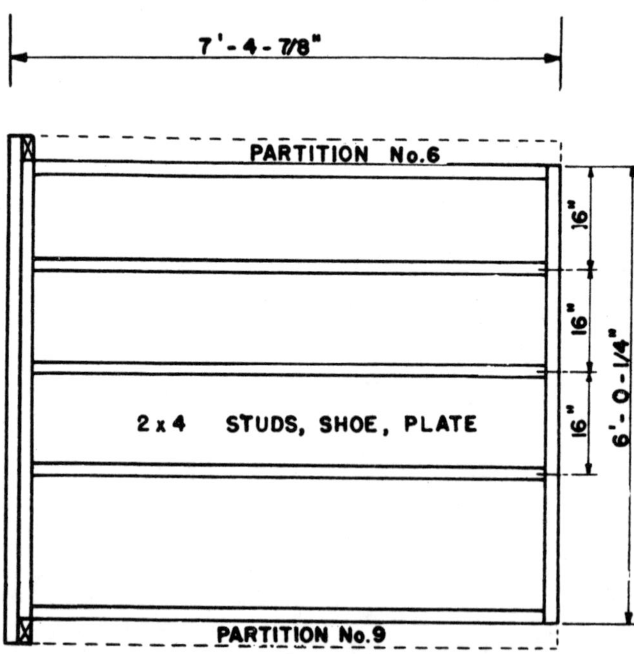

PARTITION No. 7

70

BUILD PARTITIONS

Use 2 x 4 — 4/16,23/14,7/12,2/10.
2 x 6 — 1/3.

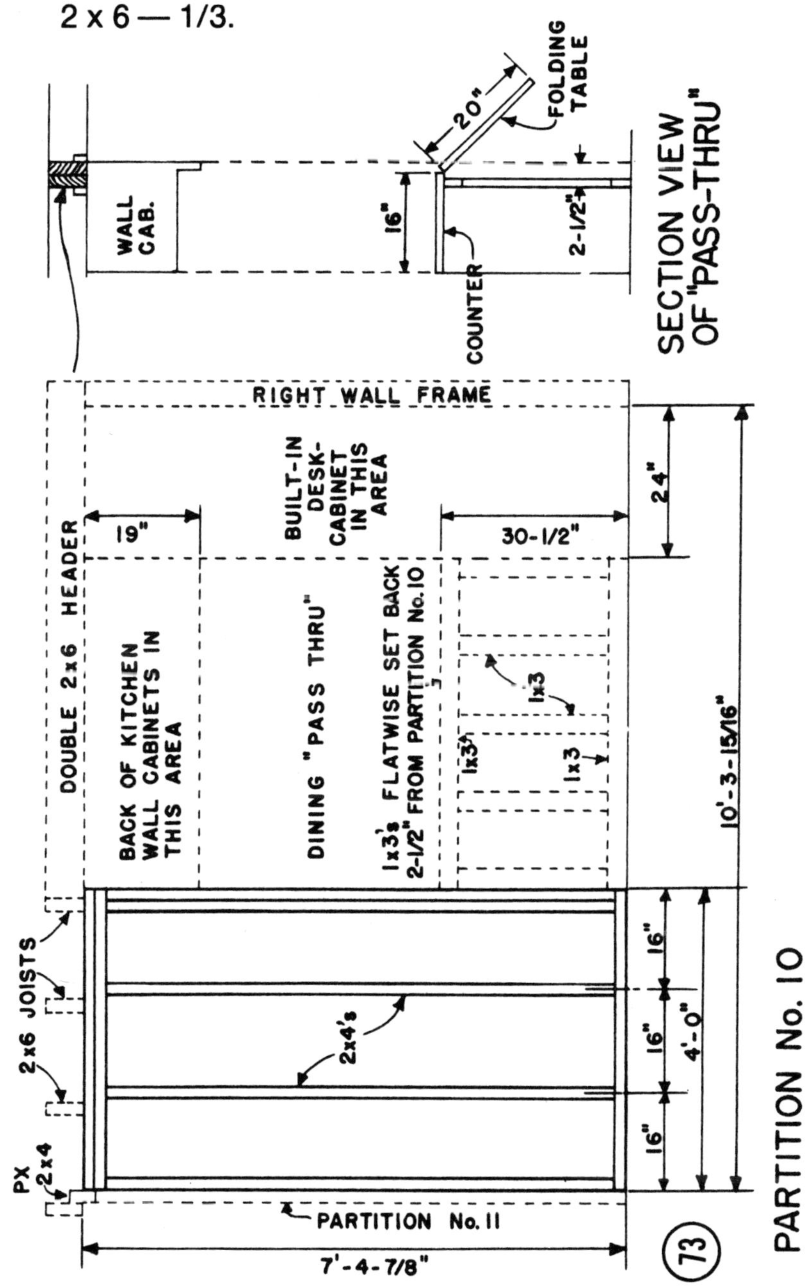

Build partitions No. 6,7,8,9,10 to height required, Illus. 69,70,71,72,73,74,75,76 and raise in sequence numbers indicate. Partitions No. 11,12,13, Illus. 74,75,76, are installed after ceiling joists are in place.

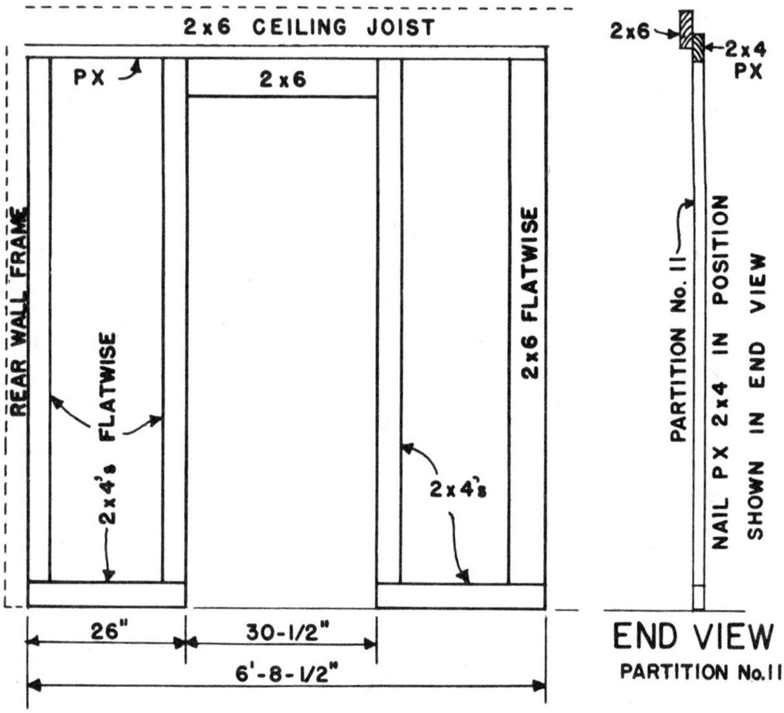

(74) PARTITION No. 11

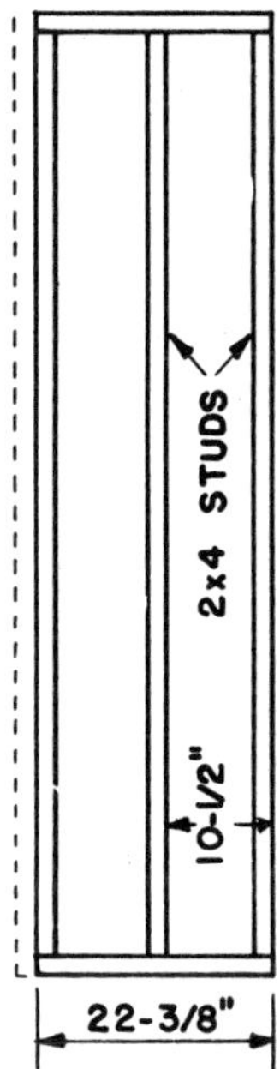

PART. No.12

(75)

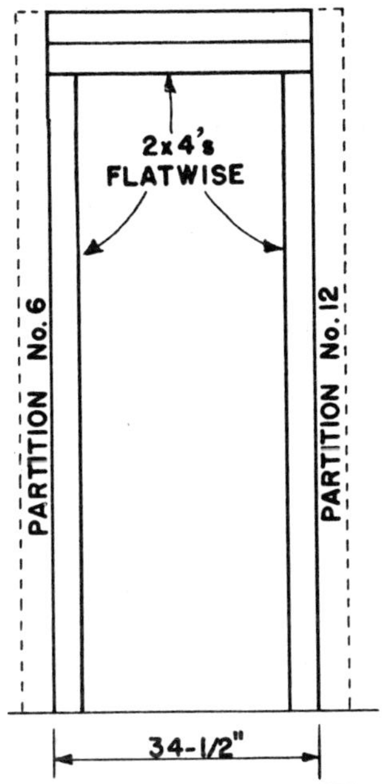

PARTITION No.13

(76)

CEILING JOISTS, HEADERS

Use 2 x 6 — 10/16,21/14,3/10,2/8.
2 x 8 — 2/10.
1 x 2 — 30 lineal feet.

Illus. 77 shows location of ceiling joists, framing for chimney, recessed lights, overhead stair opening, 2 x 6 headers, 1 x 2 ceiling joist nailors and nailors for ceiling panels, cabinets, etc.

Notch ends of ceiling joists to receive 1 x 2 nailor strips as shown in Illus. 77,78.

Cut and nail 2 x 6 headers to length required. Nail headers in position shown, Illus. 77.

Nail 1 x 2 in position, Illus. 78.

CEILING JOISTS

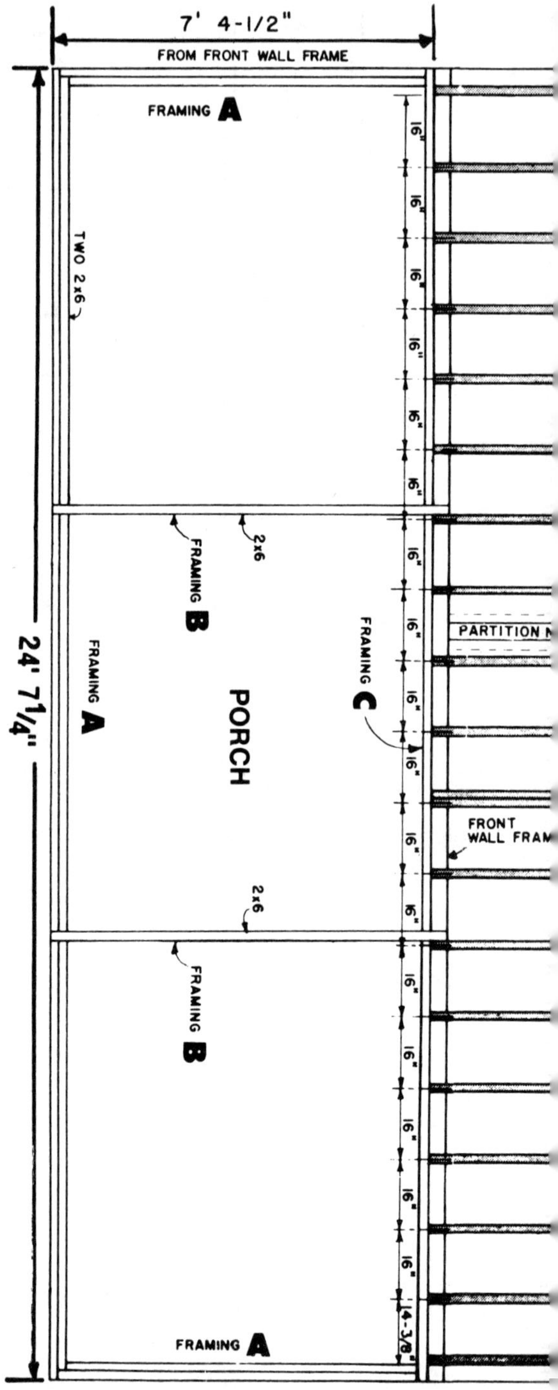

77

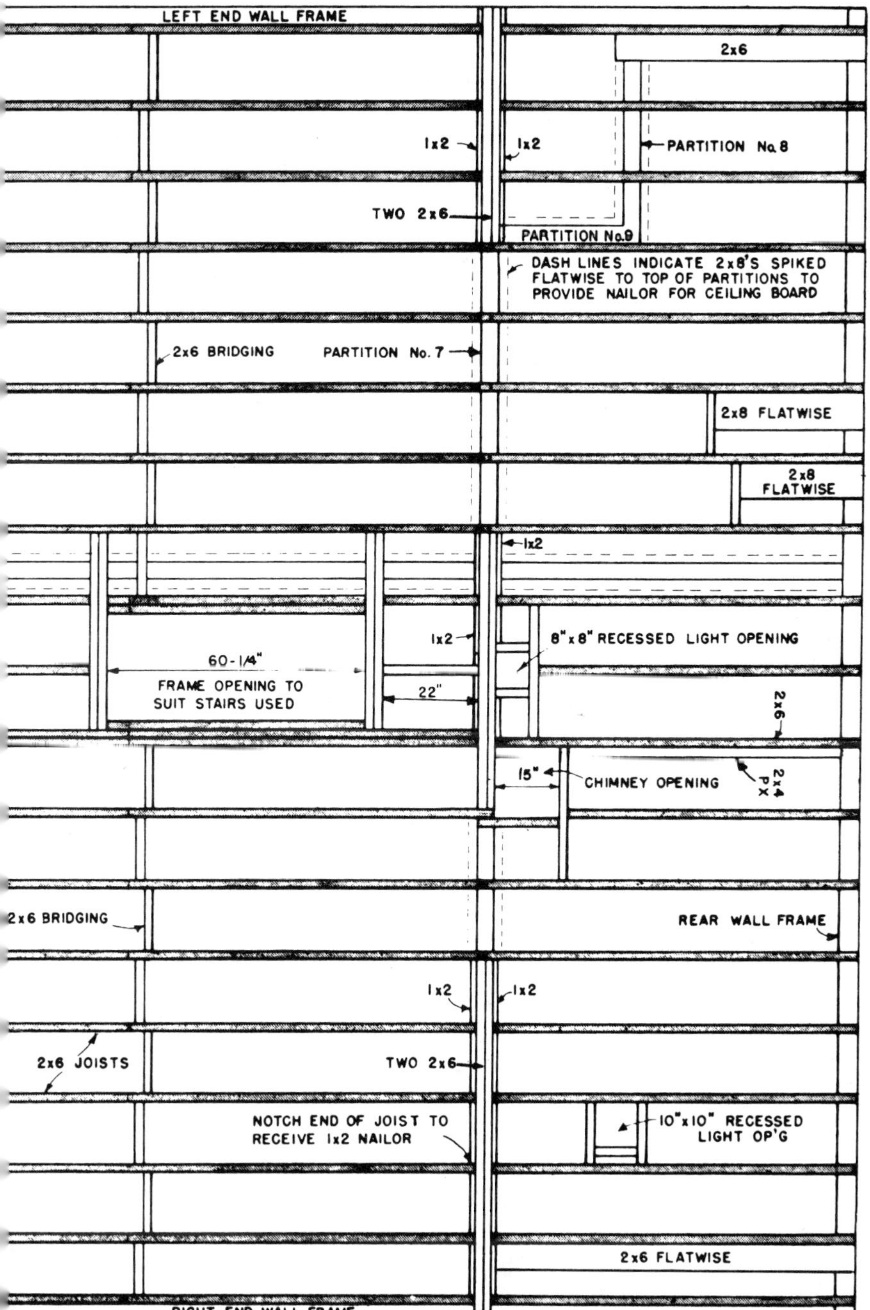
LEFT END WALL FRAME
2x6
1x2
1x2
PARTITION No. 8
TWO 2x6
PARTITION No. 9
DASH LINES INDICATE 2x8'S SPIKED FLATWISE TO TOP OF PARTITIONS TO PROVIDE NAILOR FOR CEILING BOARD
2x6 BRIDGING
PARTITION No. 7
2x8 FLATWISE
2x8 FLATWISE
1x2
1x2
8"x8" RECESSED LIGHT OPENING
60-1/4"
FRAME OPENING TO SUIT STAIRS USED
22"
2x6
15"
CHIMNEY OPENING
2x4 PX
2x6 BRIDGING
REAR WALL FRAME
1x2
1x2
2x6 JOISTS
TWO 2x6
NOTCH END OF JOIST TO RECEIVE 1x2 NAILOR
10"x10" RECESSED LIGHT OP'G
2x6 FLATWISE
RIGHT END WALL FRAME

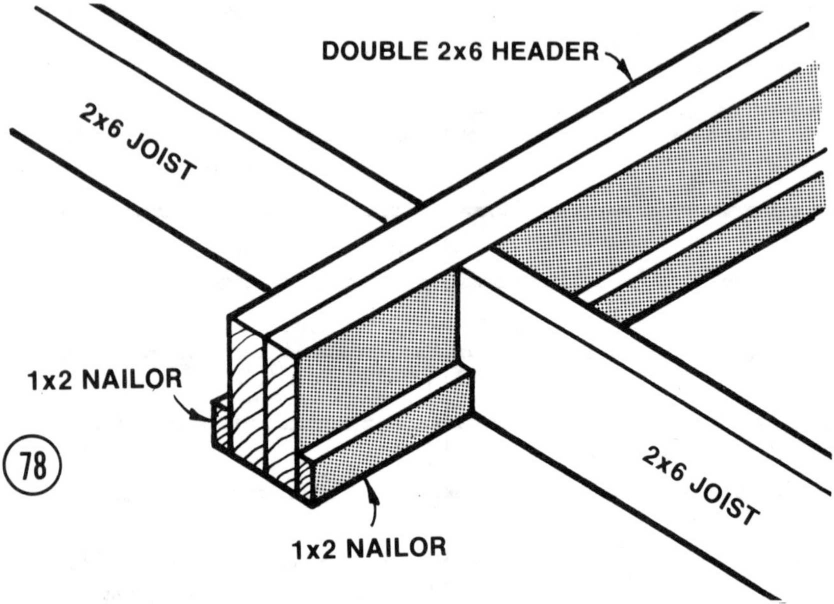

78

LAY PORCH FLOOR

To allow edge of concrete porch floor, Illus. 79, to project over foundation block, and to slope floor away from house, make a form. Place 2 x 8 A against foundation across front and side. Keep top edge of 2 x 8 flush with foundation. Nail 2 x 6 B temporarily to A. Allow B to project 2" above A.

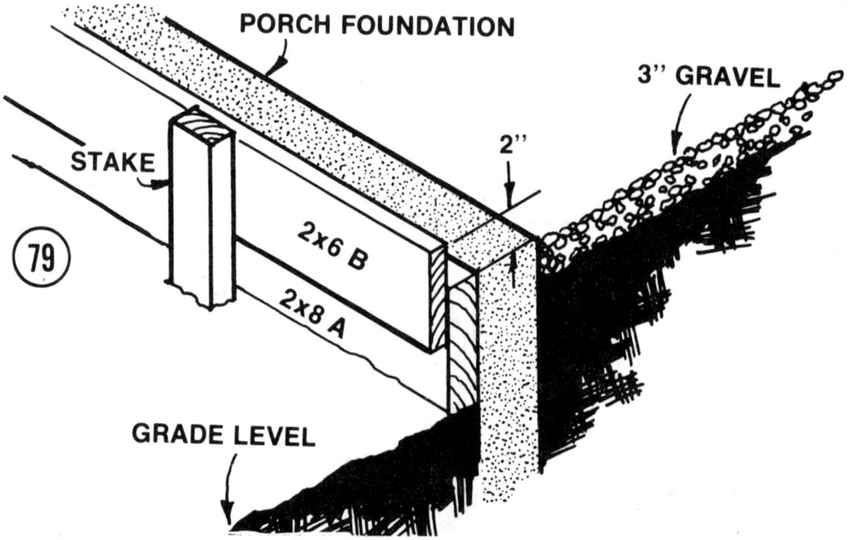

79

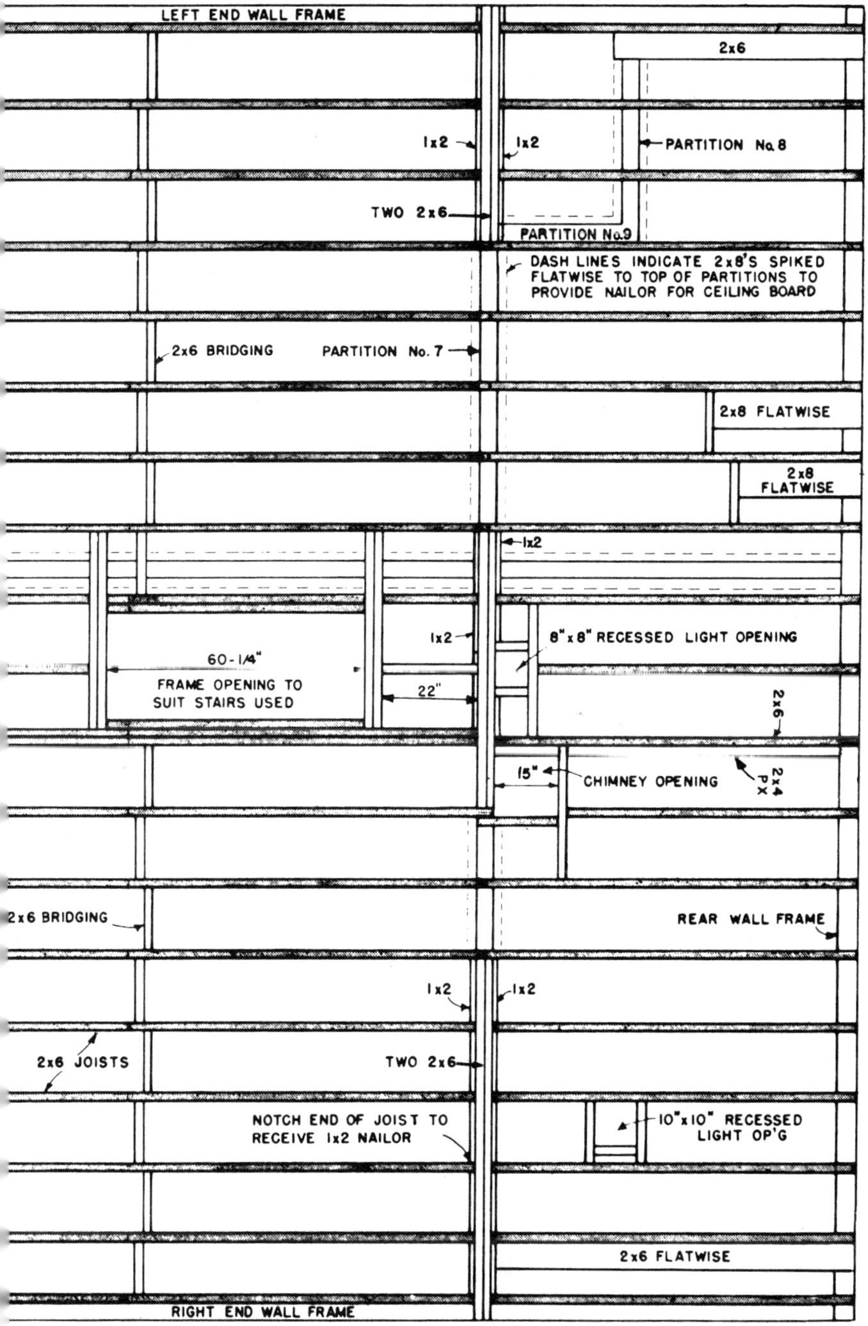
LEFT END WALL FRAME
2x6
1x2
1x2
PARTITION No. 8
TWO 2x6
PARTITION No. 9
DASH LINES INDICATE 2x8'S SPIKED FLATWISE TO TOP OF PARTITIONS TO PROVIDE NAILOR FOR CEILING BOARD
2x6 BRIDGING
PARTITION No. 7
2x8 FLATWISE
2x8 FLATWISE
1x2
1x2
8"x8" RECESSED LIGHT OPENING
60-1/4"
FRAME OPENING TO SUIT STAIRS USED
22"
2x6
15"
CHIMNEY OPENING
2x4 PX
2x6 BRIDGING
REAR WALL FRAME
1x2
1x2
2x6 JOISTS
TWO 2x6
NOTCH END OF JOIST TO RECEIVE 1x2 NAILOR
10"x10" RECESSED LIGHT OP'G
2x6 FLATWISE
RIGHT END WALL FRAME

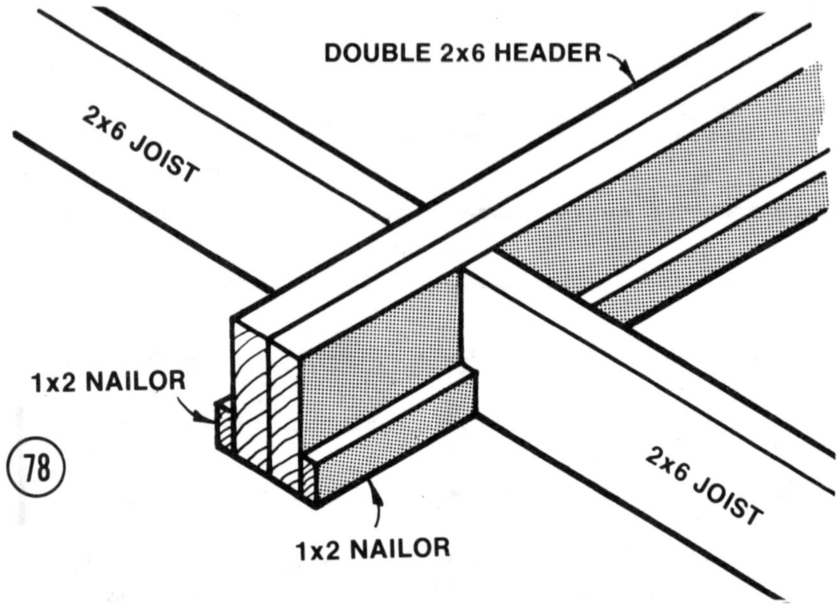

LAY PORCH FLOOR

To allow edge of concrete porch floor, Illus. 79, to project over foundation block, and to slope floor away from house, make a form. Place 2 x 8 A against foundation across front and side. Keep top edge of 2 x 8 flush with foundation. Nail 2 x 6 B temporarily to A. Allow B to project 2" above A.

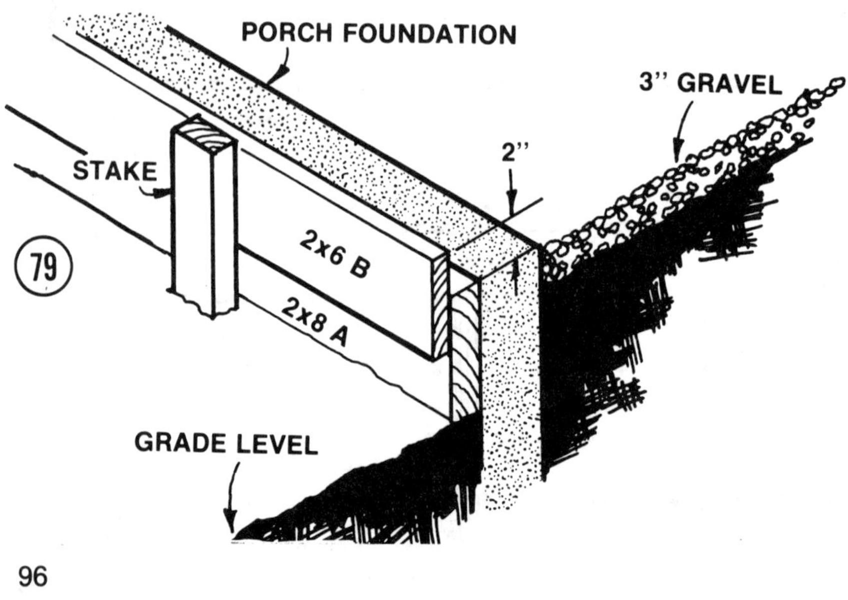

Cover front bedplate and sill beam with a strip of copper flashing.

Spread and tamp a 3'' layer of ¾'' crushed stone. Place 2 x 3's in position across front of house. Top of 2 x 3 will now be slightly higher than edge of 2 x 6 across front of porch. This represents slope of finished porch floor.

To reinforce entire edge of porch, embed an 18'' wide strip of metal lath in concrete. Place this strip about 1'' in from B.

Pour floor using 1 part cement, 2 parts sand to 3 parts of ¾'' gravel. Use a 2 x 4 as a screed to smooth and slope concrete. Before concrete has a chance to set, place planks across floor, remove 2 x 3 against house and fill void with concrete. Cover with slate, flagstone or quarry tile.

Book #617 Concrete Work Simplified explains how to score and finish a concrete deck.

FRAME PORCH

Use 2 x 6 — 4/16,2/14,3/12.
2 x 4 — 4/18,6/16.

Nail two 2 x 4's together for a porch post, Illus. 57,77,80. To estimate exact length required for each post, nail a temporary 2 x 6 x 10' to outside ceiling joists, allowing 2 x 6 to project about 8' beyond front wall frame. Check with level. Stretch a line between. The porch post equals distance from finished concrete porch deck to bottom of 2 x 6 porch plate A, Illus. 80.

Drill hole in bottom of each post to receive 1'' galvanized pipe, Illus. 44. Set porch posts in position and brace each to house and with 1 x 2 cross bracing. Check each post to make certain it's plumb. Cut each post to exact height required.

Remove temporary 2 x 6 x 10'. Nail double 2 x 6 framing ''A'' to length required to maintain overall dimension shown in Illus. 77,80.

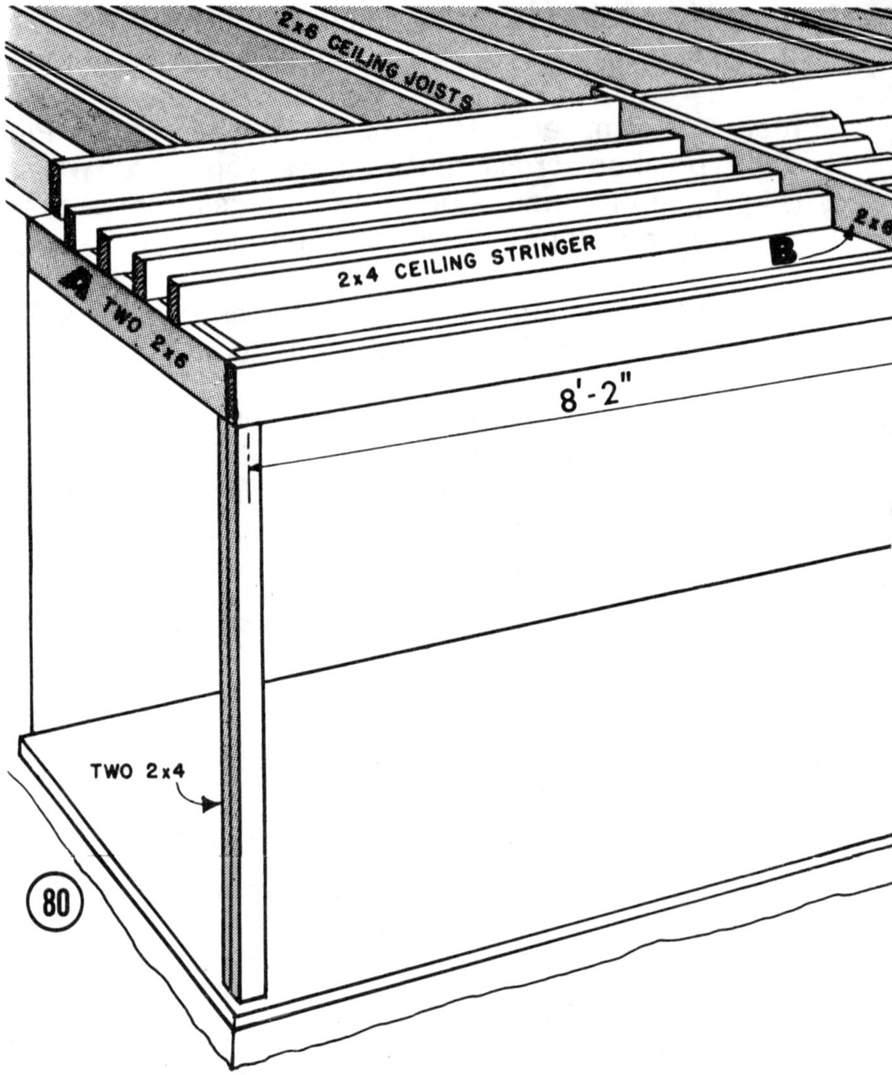

RAISE RIDGE AND RAFTERS

Use 2 x 6 — 40/16, 1/8 rafters.
    2 x 8 — 1/14, 1/12 ridge.
    2 x 4 — 6/12 for cats, or use scrap.

Cut 2 x 6 rafters to length and angle shown, Illus. 81,82.

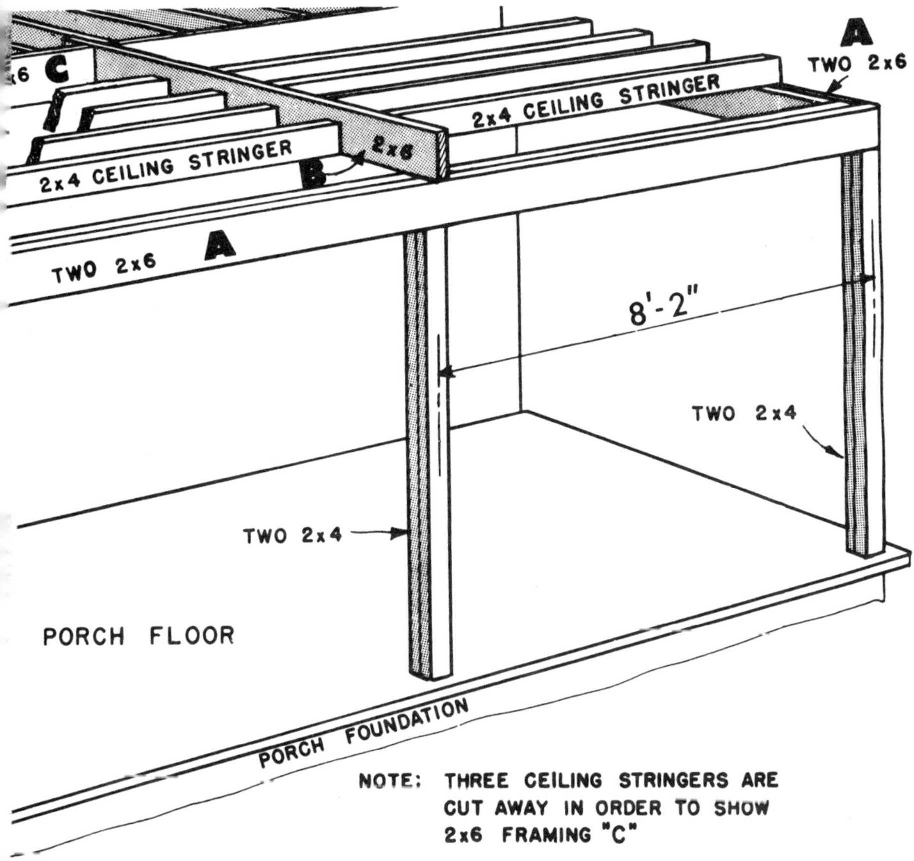

2 x 6 porch ceiling joists "B", Illus. 77,80, are nailed to top of "A" and to plate on front wall frame.

Nail 2 x 4 ceiling stringers in position indicated, Illus. 80. These stringers provide nailors for porch ceiling. Keep bottom of stringer flush with bottom of "B." Nail 2 x 6 "C" in position shown, Illus. 80.

RIDGE RAFTERS

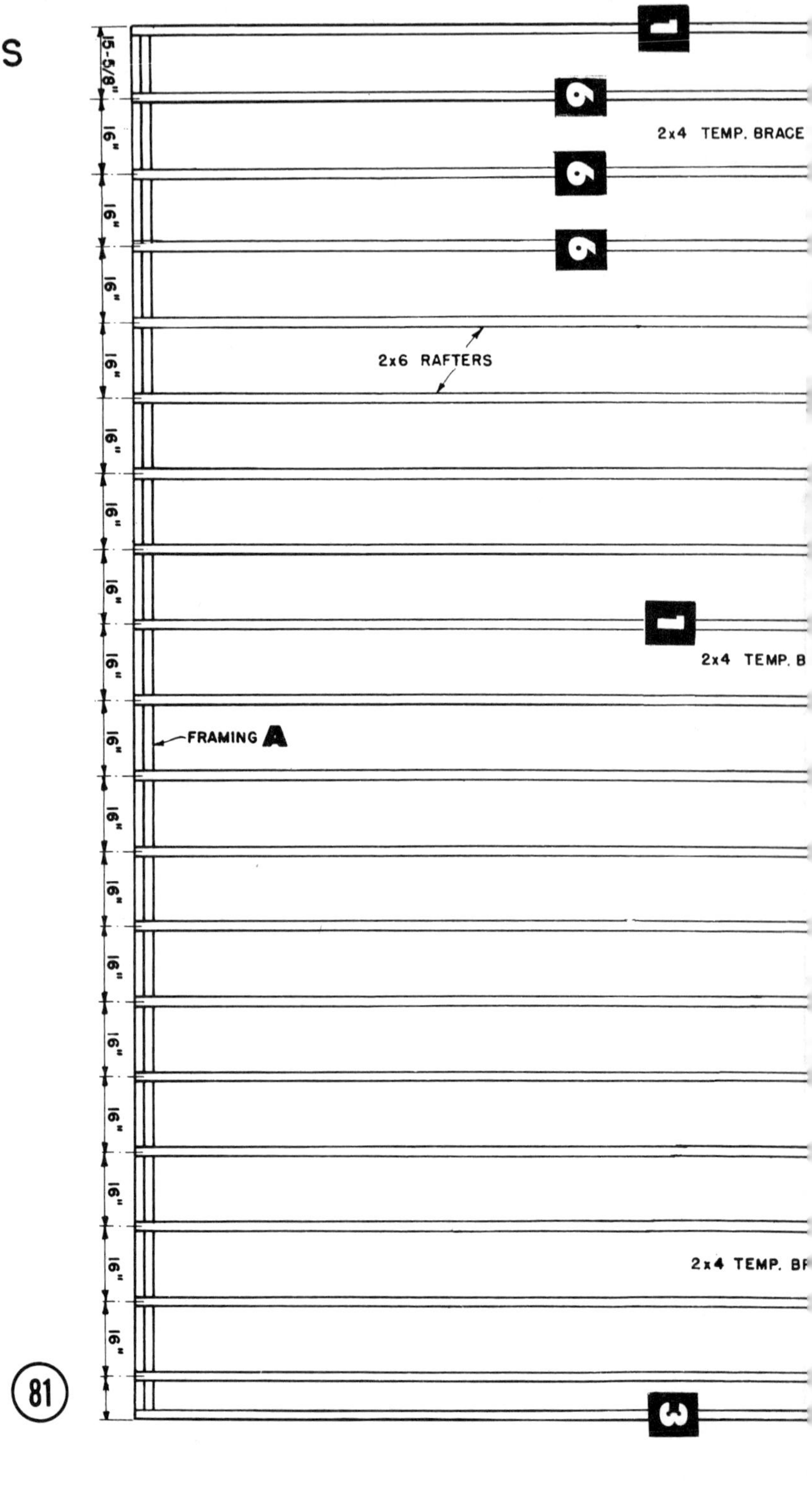

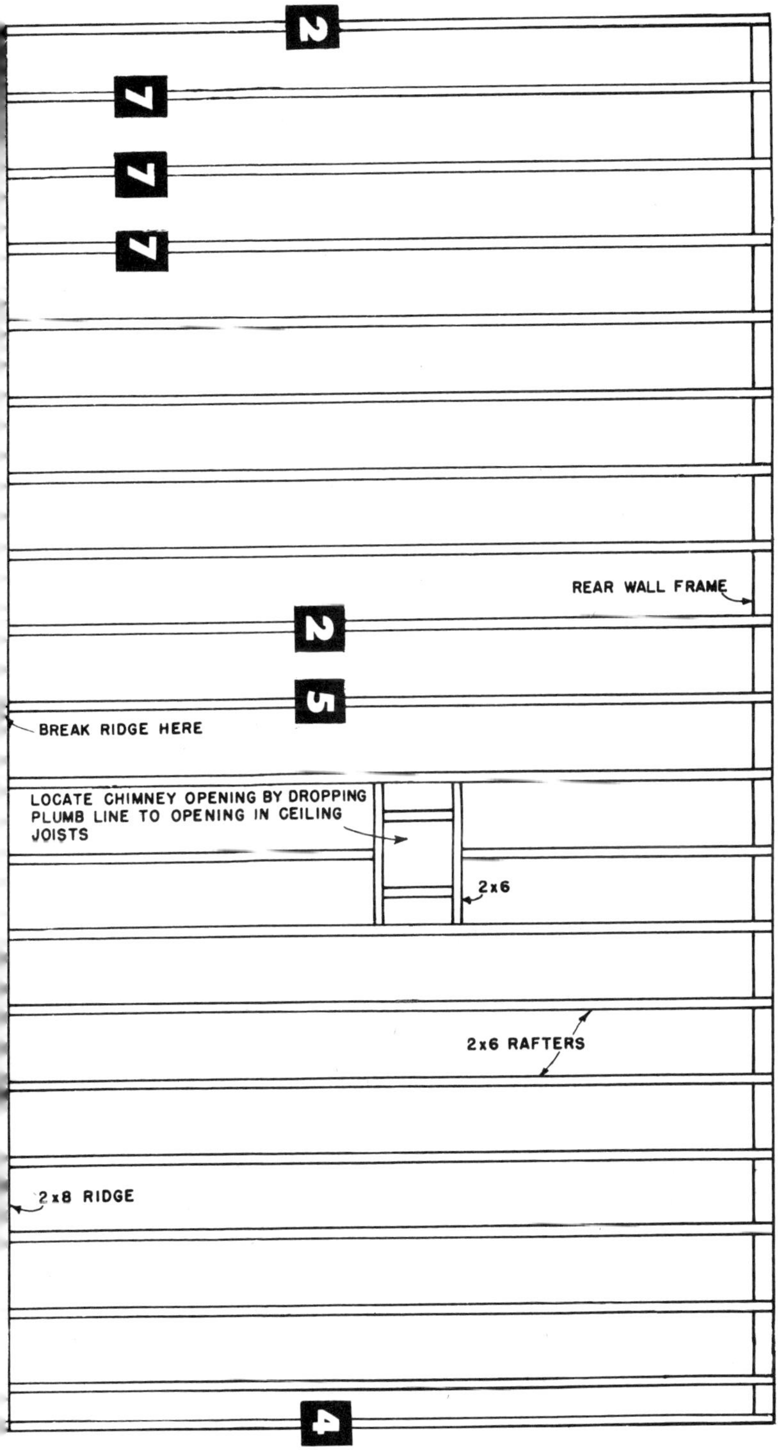
2
7
7
7
REAR WALL FRAME
2
5
BREAK RIDGE HERE
LOCATE CHIMNEY OPENING BY DROPPING PLUMB LINE TO OPENING IN CEILING JOISTS
2x6
2x6 RAFTERS
2x8 RIDGE
4

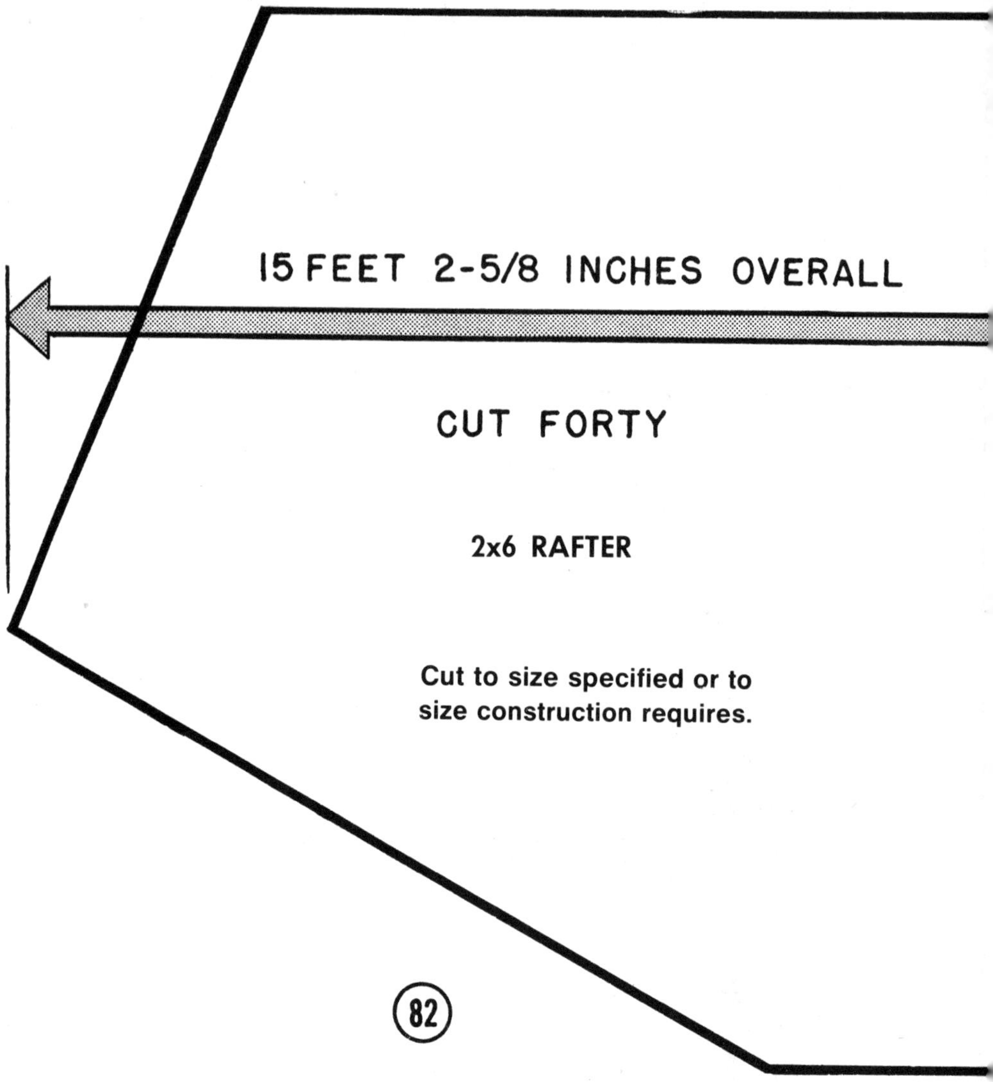

Illus. 82 provides full size pattern for cutting end of rafters. Using carbon paper, trace full size angle on cardboard or copper flashing. After cutting to exact shape, use this pattern to trace angle on one end of a 2 x 6. Measure 15'2⅝" or length required and trace other angle in position shown.

To compensate for any variation in construction, cut a pair of rafters, one #1, one #2. Use a short piece of 2 x 8 as a ridge.

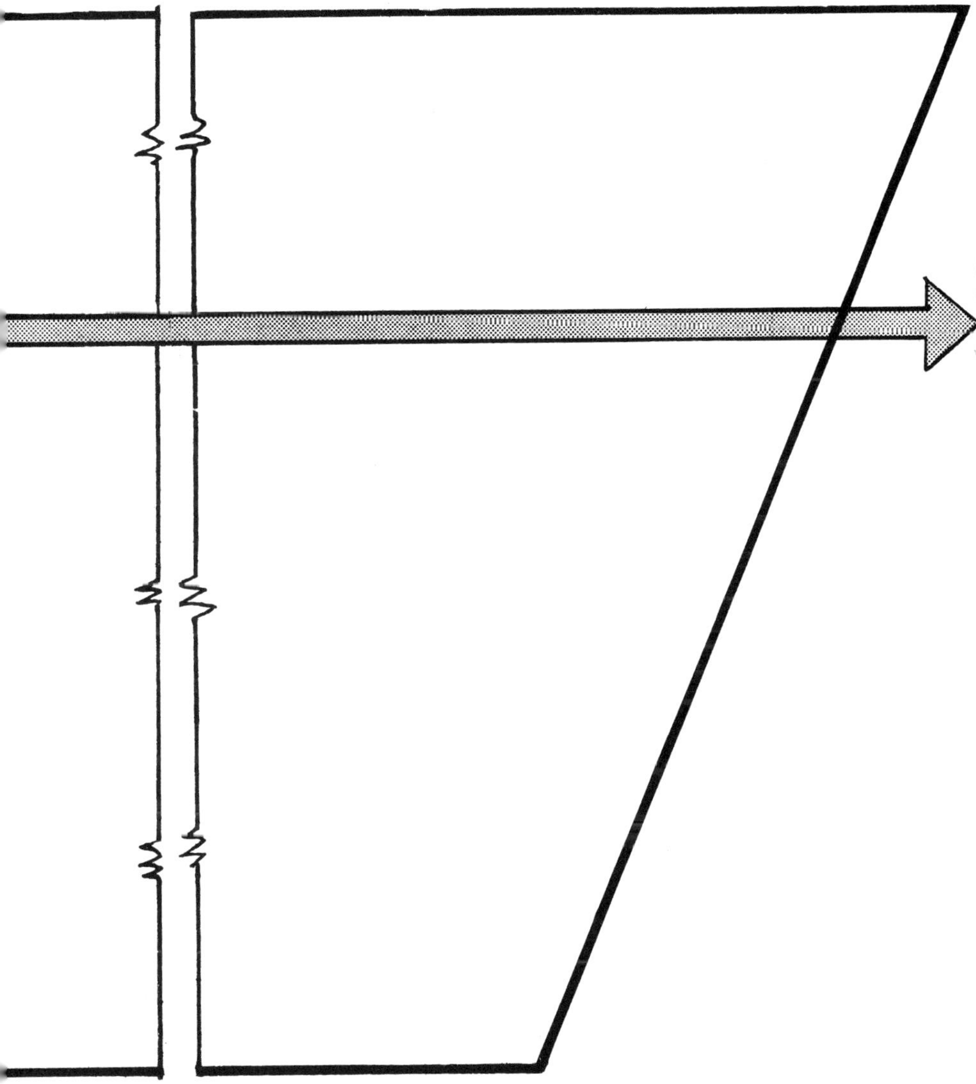

Place rafters in position. If OK, cut all rafters to size specified or size required.

To simplify flooring attic, and raising ridge and rafters, nail plywood panels to top of ceiling joists. This provides a working platform.

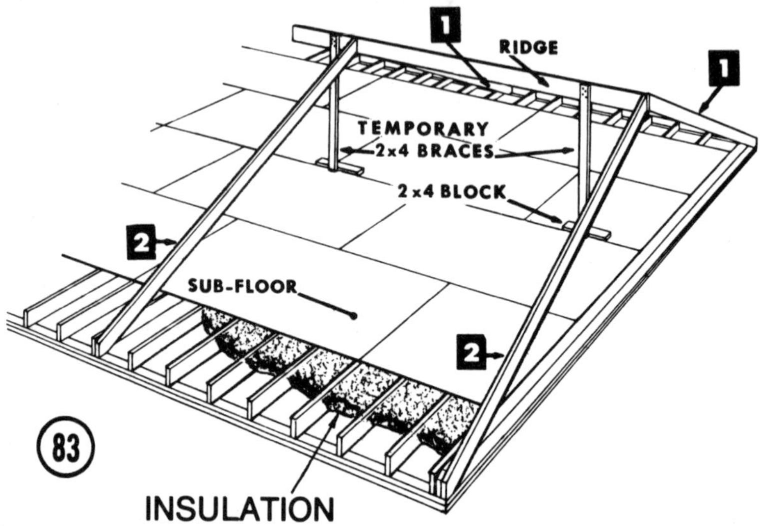

Select straight 2 x 8's for ridge. Check ends with square before cutting to exact length required. Sight up each rafter before nailing in position to make certain high point is on top.

Mark location of rafter on both sides of one length of ridge, also on both plates. Nail rafters #1 to ridge. Raise and brace ridge in position. Nail rafters to plate. Nail rafter #2 to ridge. Check ridge with level. Readjust 2 x 4 temporary bracing to hold ridge level, Illus. 83,81.

Nail rafter #3 and #4 to ridge. Raise and nail other rafters to ridge and plate. Toenail ridge together. Brace ridge in level position.

To plumb gable ends, drive a nail in end of ridge and tie a plumb bob line to nail, Illus. 54. When point of bob measures same distance from sill beam as line does from end of ridge, the ridge is considered plumb. Brace ridge in plumb position.

To make certain rafters are 16" on center, nail a 1 x 4 temporary brace across rafters, Illus. 57, approximately 8'6" from plate. Sight down and space each rafter, then hold in position with 1 x 4.

Use a plumb bob to measure and mark opening for prefabricated chimney, Illus. 81,57. Frame opening between rafters using 2 x 6.

Frame chimney opening between ceiling joists with 2 x 6 in position indicated, Illus. 77.

Frame opening for chimney between floor joists using 2 x 8 in position shown, Illus. 60.

Chop or saw ends of ceiling joists flush with top edge of rafters, Illus. 57.

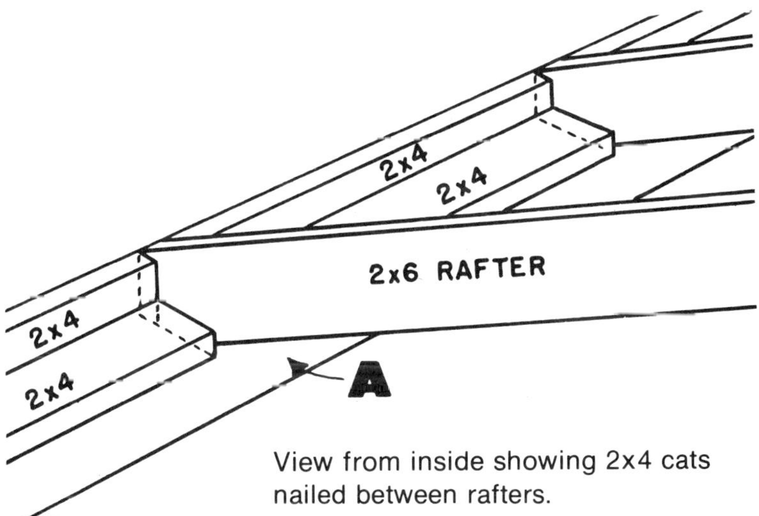

View from inside showing 2x4 cats nailed between rafters.

84

Cut and nail 2 x 4 cats between ends of rafters, Illus. 84.

If you have incorporated the porch area as part of the house, double up 2 x 6 B and nail 1 x 2 to sides, Illus. 78,80. Use 2 x 6 for ceiling joist where pitch of rafter permits. Chop ends of B where roof sheathing requires same. Or frame ceiling joists in position as shown in Illus. 148.

If you have included porch area as part of the main house and plan on adding a full second story, see pages 160,161.

GABLE STUDS

Use 2 x 4 — 8/14, 1/12.

Cut top end of gable stud to angle of pattern shown full size, Illus. 85, and to length required to fit in position noted in Illus. 86. Toenail to plate and rafter with 8 penny nails.

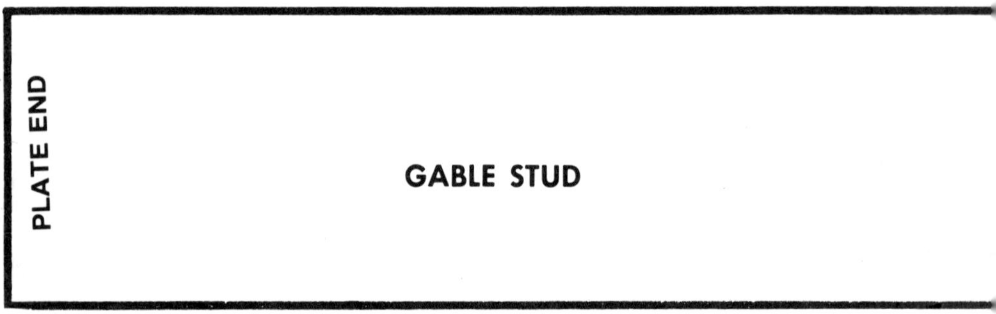

85

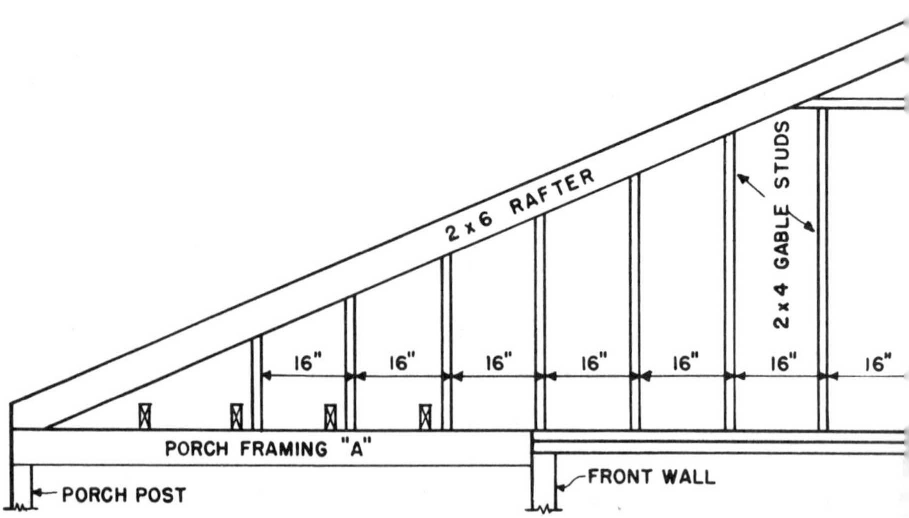

86

NOTE: Frame opening to fit aluminum louver.

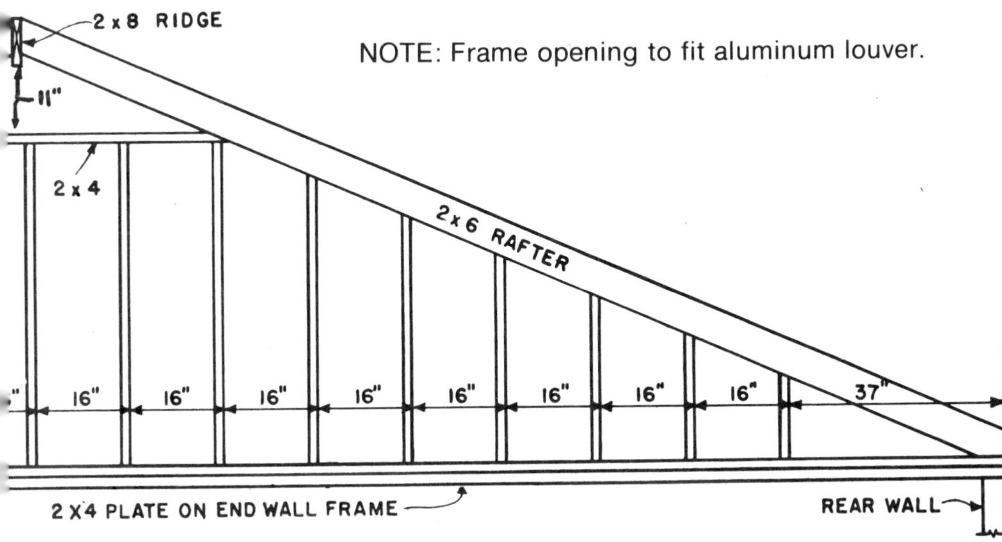

FRAME FOR LOUVER

Frame opening for adjustable aluminum louver, Illus. 86. We indicate 11" opening. Cut ends of header to angle rafter requires. Check with level and toenail to rafters, Illus. 87.

Louver is installed after ⅜" sheathing has been applied.

Chop off bottom edge of ridge to fit louver.

87

RAFTER STUDS

Use 2 x 4 — 1/14, 10/12.

Nail 2 x 4 shoe in position indicated, Illus. 57. Cut rafter studs to length required, check with level and nail to each rafter and shoe, Illus. 88.

If you move a partition, nail 2 x 6 bridging every two feet between ceiling joists. This permits nailing 2 x 4 shoe, Illus. 88, in position required.

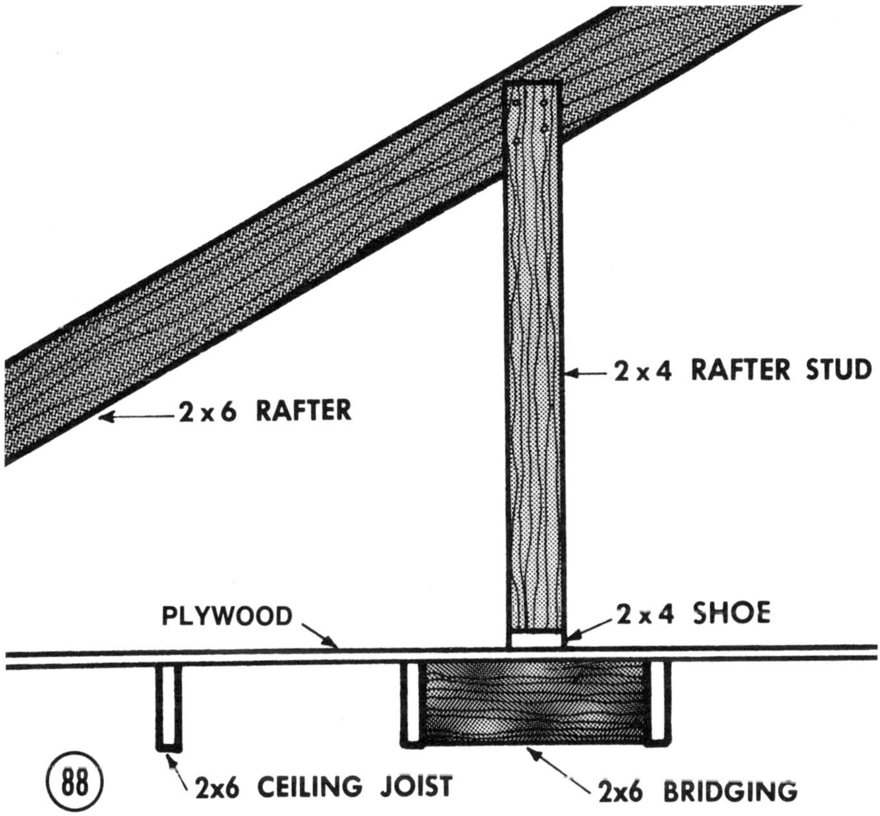

(88)

APPLY PLYWOOD SHEATHING

Approx. 25 — 3/8" x 4 x 8' plywood panels.

#15 roofing felt to cover 850 sq.ft.

Use thickness sheathing local codes or FHA recommends. We suggest 3/8" and placed bedplates 3/8" from edge of block to accommodate same, Illus. 39.

3/8" plywood sheathing butts against termite shield and is applied horizontally. Butt panels over center of studs. Always stagger vertical joints so adjoining panels don't butt together

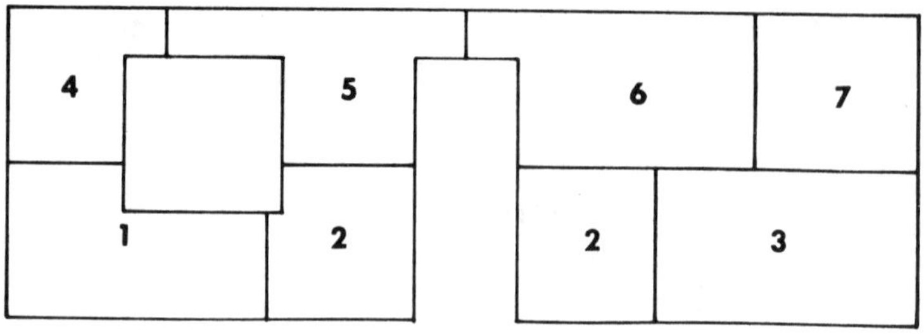

Front Wall

89

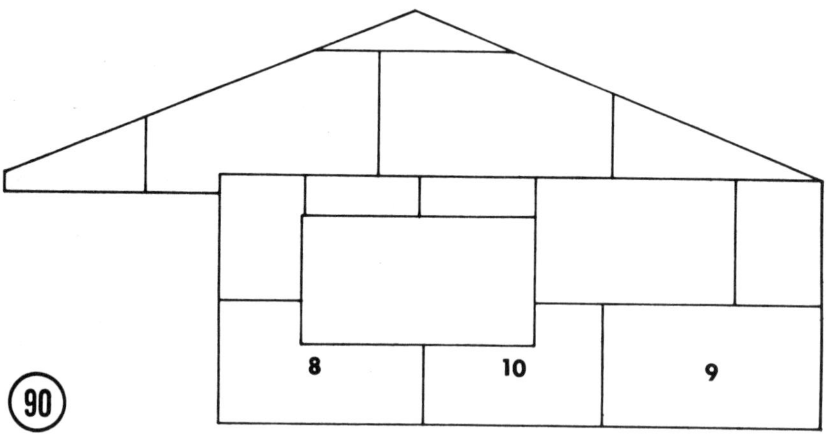

90

Right Wall

over same stud. Illus. 89 shows sheathing on front wall frame. Illus. 90 shows sheathing on right wall frame.

Cut plywood flush with top edge of rafter. Use 8 penny common nails every 6'' along edge of panel, every 12'' on inner studs.

Remove temporary bracing and nail 1 x 4 wind braces in position outlined on page 82.

Apply ⅜'' plywood to the porch posts where indicated, Illus. 91.

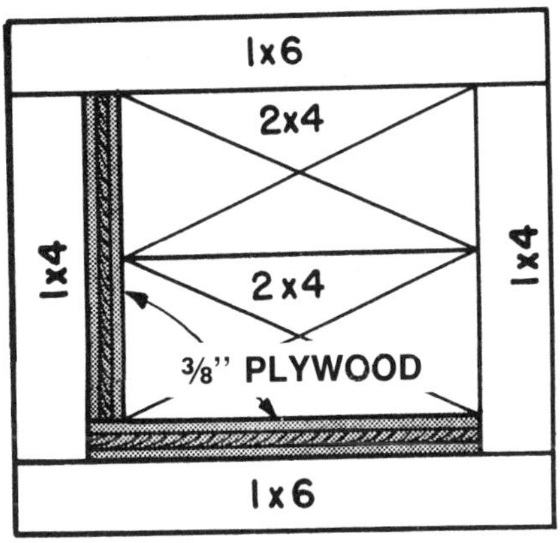

LEFT END POST

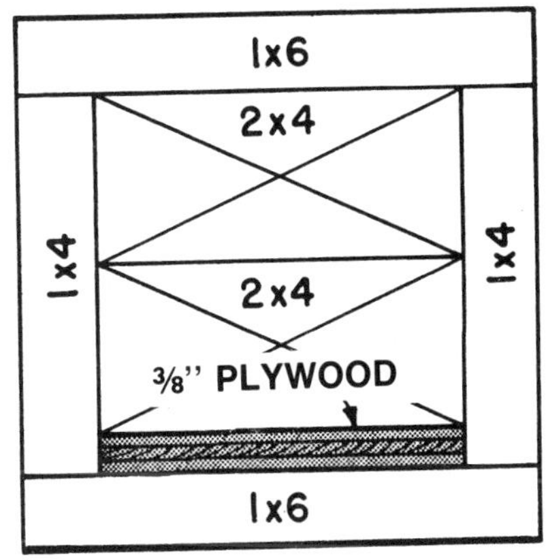

MIDDLE POSTS

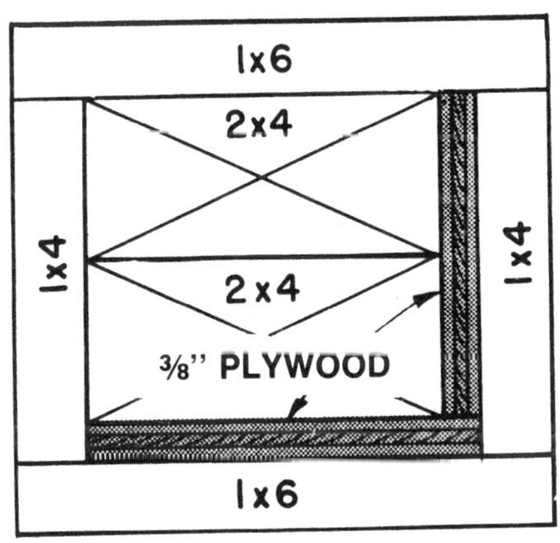

RIGHT END POST

(91)

Apply 3/8" plywood to porch header "A," Illus. 92; to face of rafters, Illus. 93, and to gable ends, Illus. 94.

As there are different styles of window and door frames, your dealer can advise whether frames are to be nailed after plywood sheathing has been applied, or directly to studs. If frames are applied to studs, set window and door frames before applying plywood sheathing.

3/8" exterior plywood can be nailed directly to ceiling josts on porch.

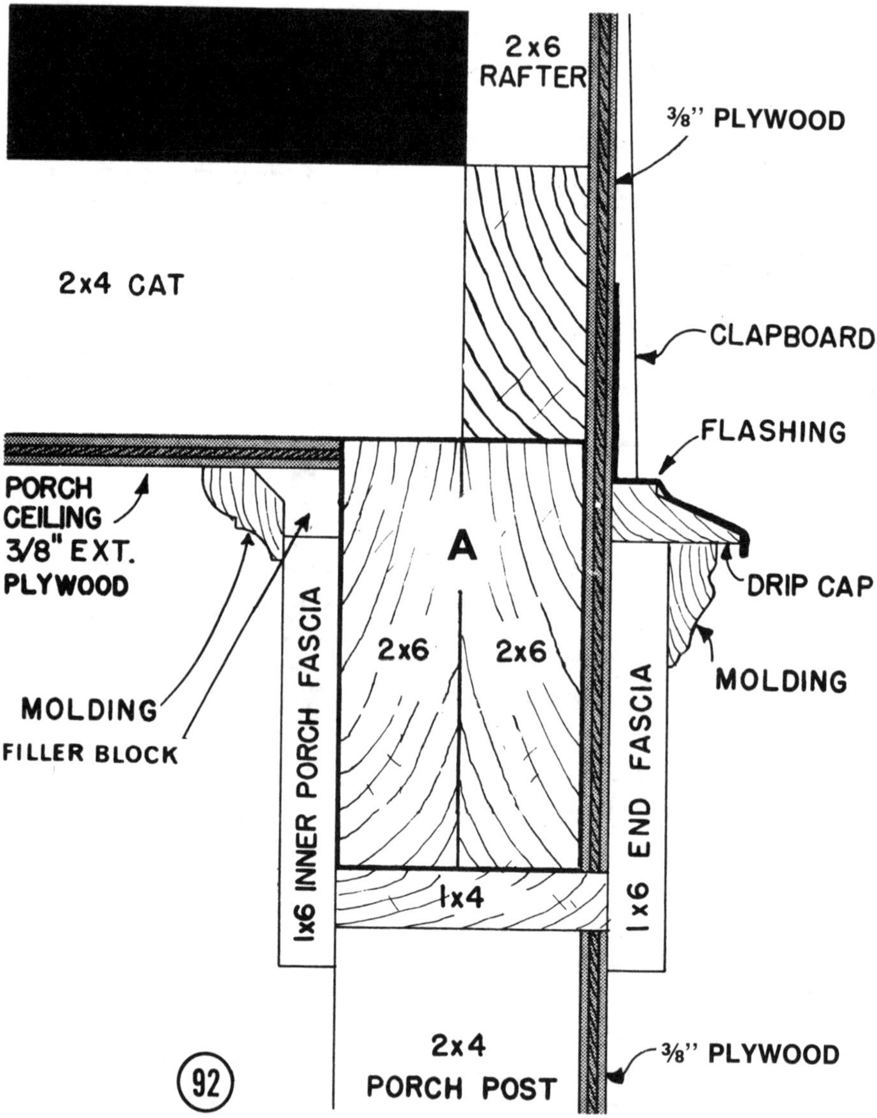
2x6
RAFTER
3/8" PLYWOOD
2x4 CAT
CLAPBOARD
FLASHING
PORCH
CEILING
3/8" EXT.
PLYWOOD
A
DRIP CAP
2x6
2x6
1x6 INNER PORCH FASCIA
1x6 END FASCIA
MOLDING
MOLDING
FILLER BLOCK
1x4
2x4
PORCH POST
3/8" PLYWOOD
92

PORCH END

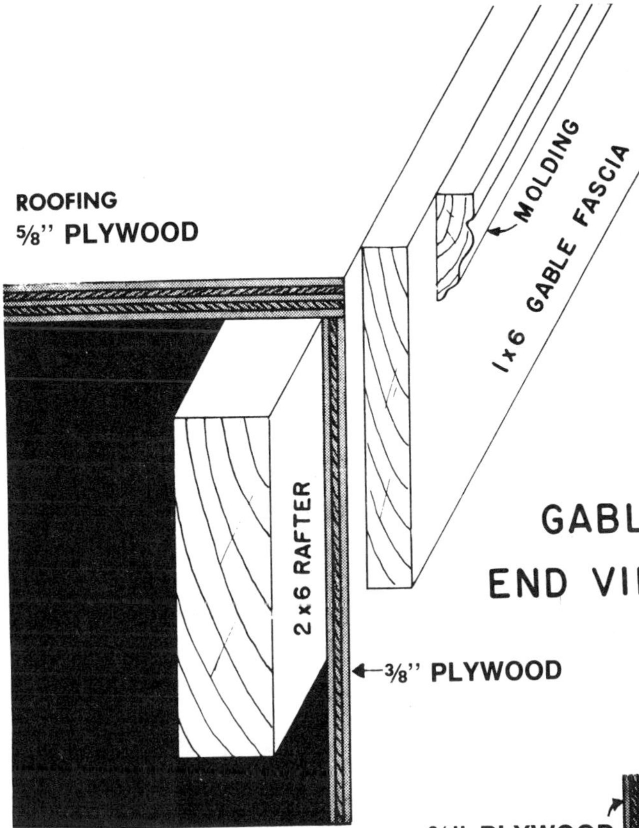

GABLE FASCIA
END VIEW

93

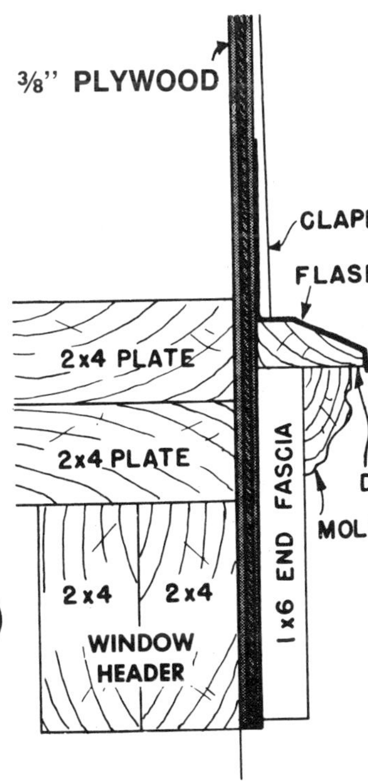

GABLE END

94

ROOFING

Build scaffold to height required, Illus. 95,96. Use 2 x 4 posts, 1 x 6 cross bracing. Nail 2 x 4 blocks temporarily to studs. Nail cross bracing to side of house and to blocks. Make certain posts are plumb. If ground is soft, place each post on a piece of 2 x 6 or 2 x 8. Use 2 x 6 or 2 x 8 for platform. When removing scaffold, fill all nail holes with putty.

95

A–2 x 4

B–2 x 4

C–2 x 6 or 2 x 8

D–1 x 6

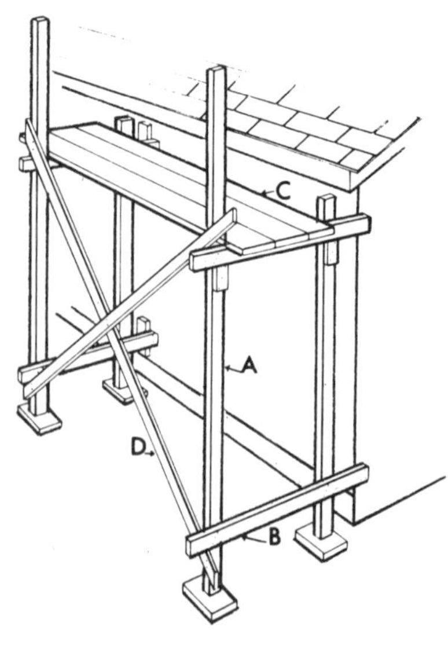

96

Before nailing plywood sheathing to rafters, place all the 5/8" plywood needed for attic flooring on top of ceiling joists. Nail panels in place to provide a working platform.

Before applying plywood to ceiling joists, Illus. 57, frame opening to size overhead stair manufacturer recommends. Illus. 77 shows headers recommended for most overhead stairs. Install pull-down attic stairs. Your dealer can advise exact size of rough opening required for stairs he sells.

Apply thickness insulation between ceiling joists local conditions require.

97

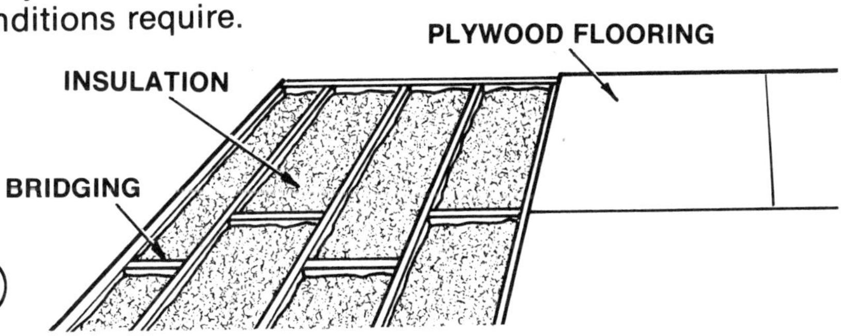

Approx. 10—5/8" x 4 x 8' plywood panels will be needed for attic flooring. Temporarily nail a few panels in place to provide a working platform. After rafters have been raised, roof sheathed and covered with #15 felt, remove these panels and insulate between ceiling joists using thickness climatic conditions require, Illus. 97.

Nail plywood to joists with 6 penny nails. Panels butt over joists; run face grain of plywood across joists, Illus. 97.

Approx. 24– 5/8" x 4 x 8' plywood panels will be needed for roof sheathing plus #15 roofing felt to cover 750 sq. ft.

Nail plywood to rafters every 6" with 8 penny nails. Apply plywood with face grain across rafters. Start at end of rafter, allowing first panel to project over end of rafter 7/8", Illus. 98. Stagger panels so ajoining panels don't butt together over same rafter. Apply roof sheathing flush with side sheathing. Remove temporary 1 x 4 rafter braces, Illus. 57.

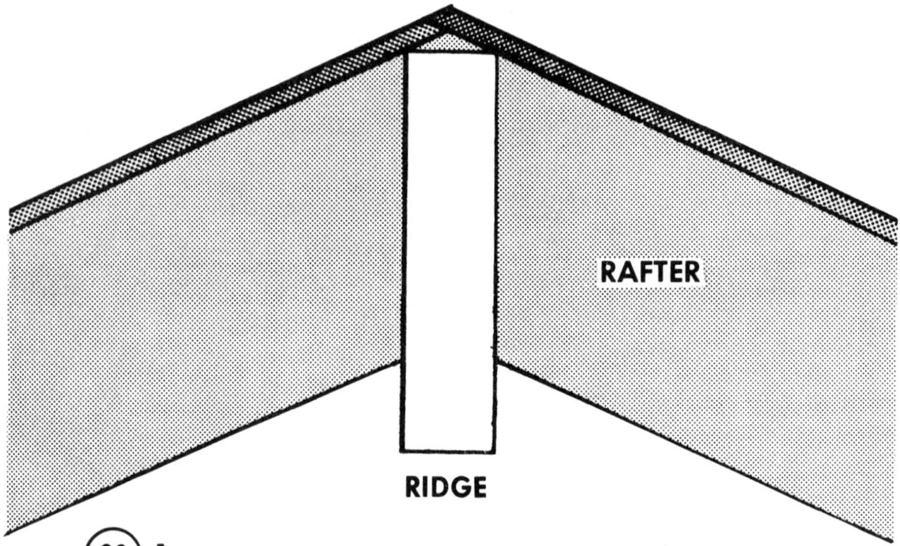

98 A

Saw panels at ridge to angle required, Illus. 98A. Start at bottom of other side and work up to ridge. Saw panels at ridge flush with other sheathing.

Staple or nail #15 felt horizontally over roof. Use big head roofing nails or stapling gun. Start at bottom and overlap each course 4 to 6". Work your way up roof, over ridge. Again start at bottom on other side and work your way up over ridge. Overlap other felt at ridge.

Allow felt to project over gable ends so it can be cut off flush with face of fascia after same has been nailed in place.

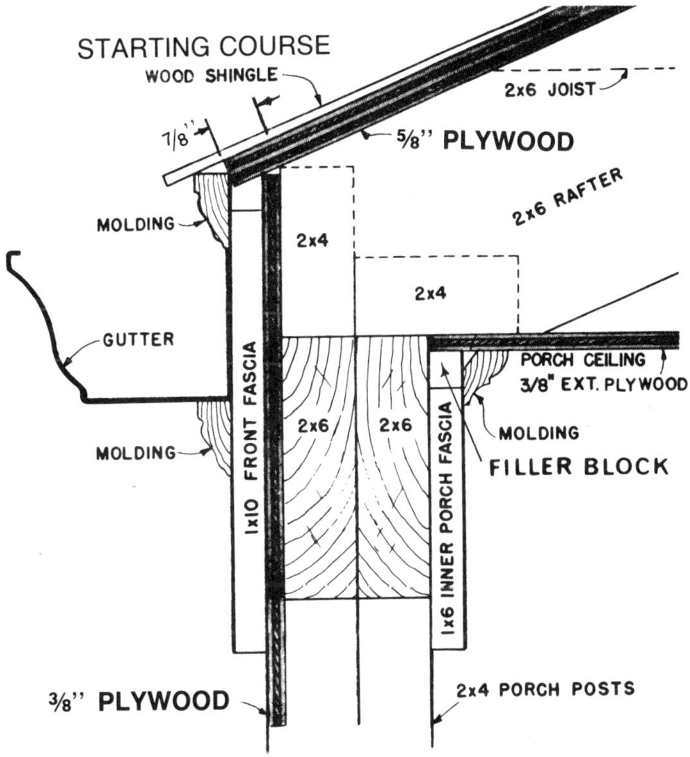

(98) FRONT FASCIA and PORCH FRAMING

NAIL FASCIA BOARDS

Use 1 x 10 — 2/14,2/12
1 x 6 — 6/16,2/14
¾ x 1⅝" Drip Cap — 2/16,2/14
6 penny aluminum nails

Note Illus. 92,93,94,95,98. Prime coat fascia boards, trim and cedar siding, if same is used, on both sides with an oil base primer before applying. Nail strips of ⅜" plywood sheathing to outside of porch framing and faces of 2 x 4 porch posts, Illus. 91, before applying trim. Cut 8" wide strips of #15 roofing felt and staple to gable ends before applying fascia, Illus. 87.

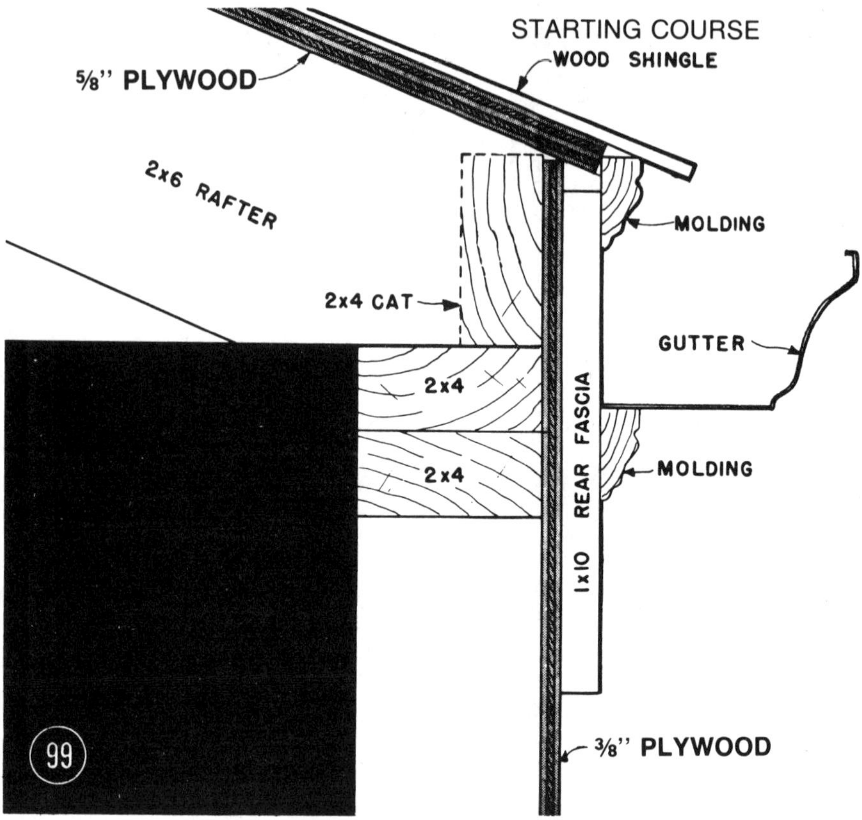

REAR FASCIA

Follow this procedure when applying fascia. Cut and nail 1 x 10 front and back fascia in position shown, Illus. 98,99.

NOTE: The bottom edge of fascia should project 3/8" below face of 1 x 4 nailed between porch posts, Illus. 92.

Stretch a line between bottom edge of front and rear fascia. Nail 1 x 6 side fascia to this line, Illus. 13,94. Miter cut fascia at corners.

Illus. 99 shows double 2 x 4 header over window.

Nail drip cap in position indicated, Illus. 92,94. Cover with copper flashing. Nail 1 x 6 gable fascia in position, Illus. 13,14,93.

Nail ¾ x 1¾'' flat shingle molding in place flush with top edge of 1 x 6 gable fascia, Illus. 93. A molding can be nailed in place along bottom of gutter, Illus. 98,99, after gutter is installed on front and rear.

CORNER BOARDS

Use 1 x 6 — 10/16,1/14
1 x 4 — 4/16,2/14,1/10
6 penny aluminum nails

Cut 1 x 6 to length and width required for corner boards, Illus. 11,12,13,14,100. Nail in position noted in Illus. 100. Nail 1 x 4 trim around windows. Window trim butts against fascia board.

Many codes permit applying ⅜'' textured plywood directly to studs. No sub-sheathing is required.

Box porch posts with 1 x 6 and 1 x 4, Illus. 91.

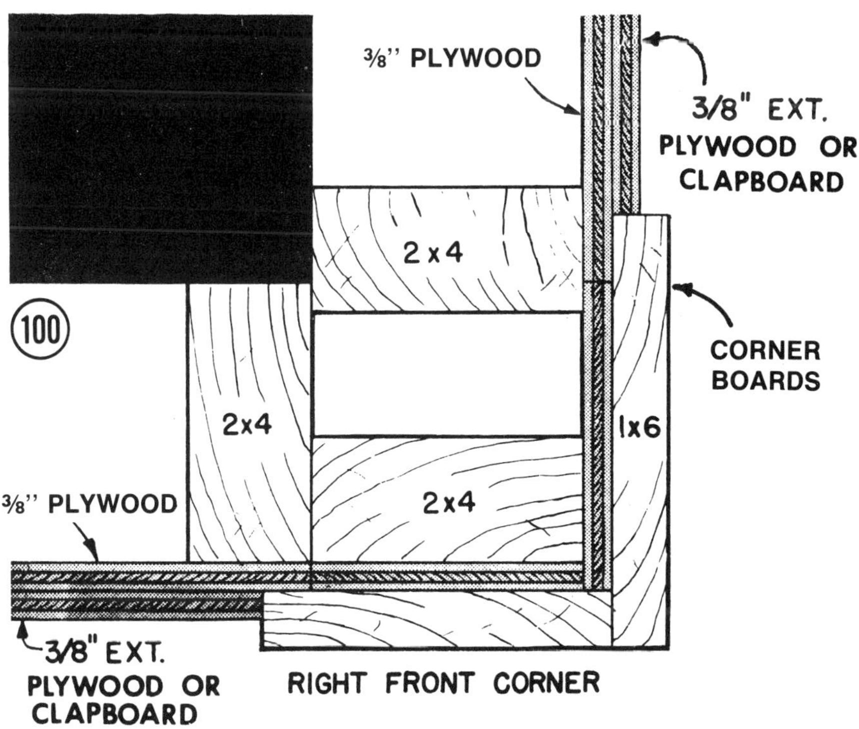

100
RIGHT FRONT CORNER

Nail 1 x 6 inner porch fascia around inside of porch flush with bottom edge of outer fascia, Illus. 98. Nail 1 x 4 between posts, Illus. 92.

APPLY ROOFING

Fiber glass, asphalt or wood shingles can now be applied or same can be applied after prefabricated chimney has been installed. Book #696 Roofing Simplified explains how a pro performs each step from applying eave, gable, valley and step flashing to application of roofing.

Shingles will be required to cover 750 sq. ft. Regardless of shingle selected, apply a starter strip in position shown, Illus. 98,99. Allow starter course to project over edge as shown. First course of roofing finishes flush with edge of starter strip. Overlap each course amount manufacturer specifies. Use care not to drive any nails through eave or gable flashing.

GUTTERS AND LEADERS

Install gutters and leaders, Illus. 98,99,101,102.

Gutters usually come in 10' lengths. Buy matching end caps, A; collectors, B; connectors, C; gutter, D; brackets, E; or other hangers F; leaders, G; leader brackets, H; elbows, K; and adhesive manufacturer recommends, Illus. 101.

The following materials will be required:

- 5 — 10' lengths of aluminum box gutter
- 4 — aluminum outlet sections
- 4 — 10' lengths of aluminum 2⅜x3¼" rectangular leader

The following attaching parts are needed:

- 8 — slip connectors
- 4 — end caps (2 left, 2 right)
- 24 — strap hangers
- 4 — aluminum strainers
- 8 — leader brackets

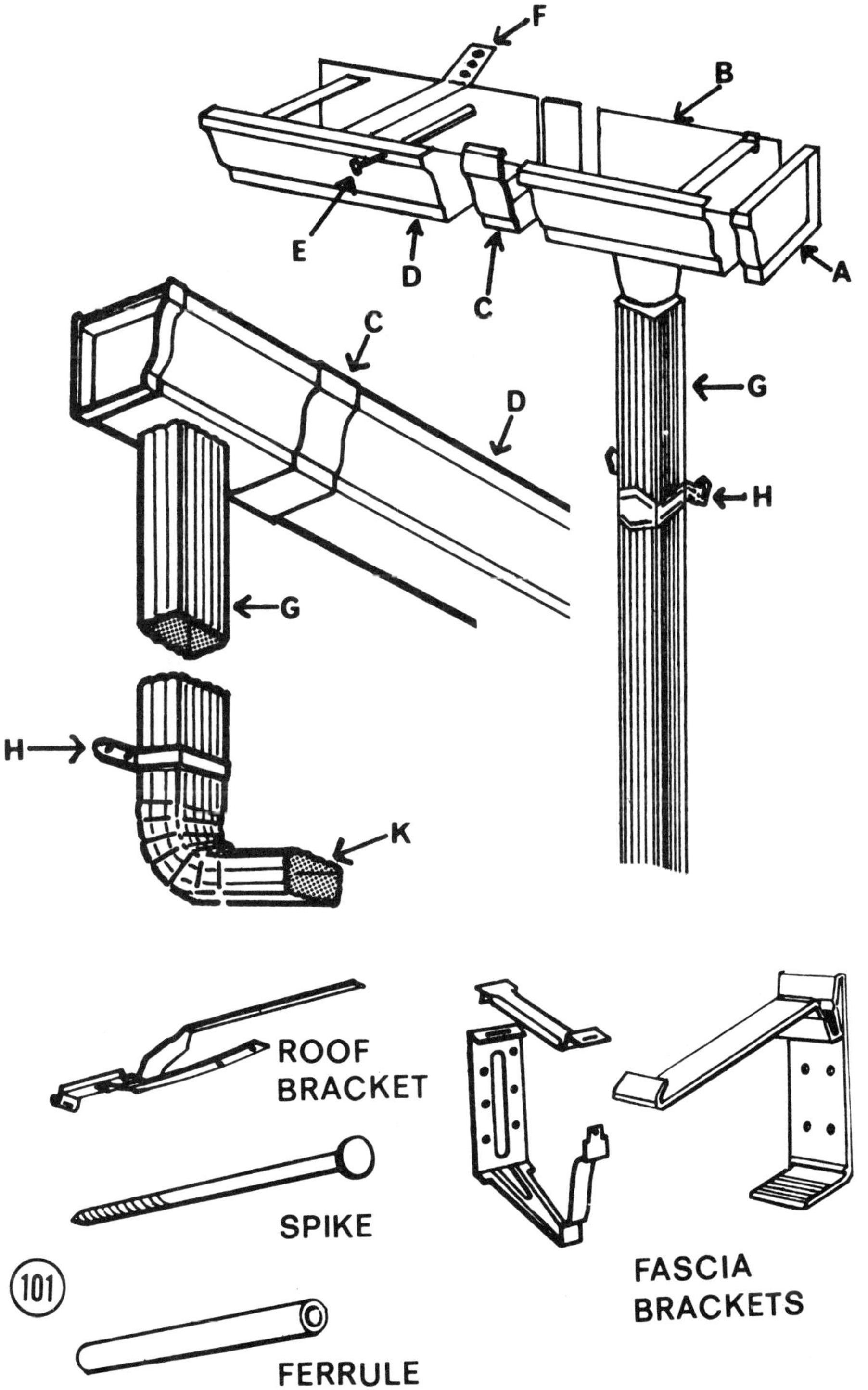
F
B
E
D
C
A
C
D
G
H
G
H
K
ROOF BRACKET
SPIKE
FERRULE
FASCIA BRACKETS

101

Aluminum nails are included with attaching parts. Follow manufacturer's directions for installing.

To obtain necessary drain off, locate and draw a line at center of fascia. Snap a line from center to each end. Slope line ¾'' from center to corner, Illus. 102. Run line from center to other corner.

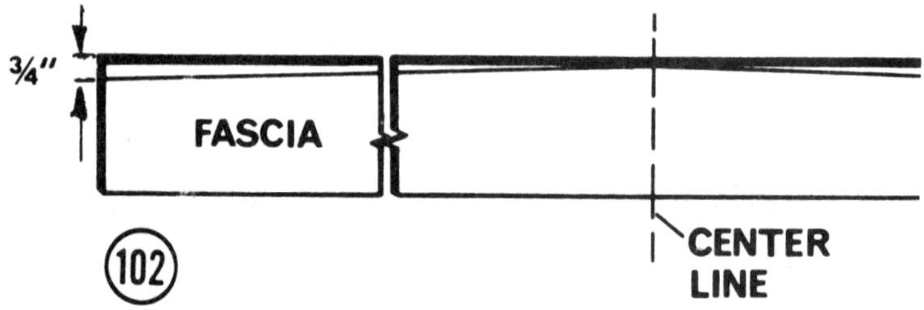

102

Preassemble end cap to collector, collector to gutter, following directions and using adhesive manufacturer provides. Many retailers now supply gutters complete with collectors. Fasten top edge of gutter to line with nails or brackets manufacturer specifies. Connect gutters at center with connector.

Fasten elbow K, Illus. 101, to bottom of leader, fasten leader to collector. Fasten leader to post with brackets manufacturer recommends. Install gutters across back following same procedure.

Make certain drain off from leader flows away from house.

INSTALL WINDOWS

Rough openings indicated are for stock size casement windows. Your dealer can mark illustration with rough opening size required for windows available. Buy windows complete with sills. Install windows following manufacturer's directions. If you install aluminum windows, be sure to use aluminum nails when nailing through aluminum.

INSTALL CHIMNEY

Manufacturers provide Underwriters Approved prefab chimneys. These contain all parts needed to install a chimney within the house in position shown or outside. Book #674 How to Install a Fireplace covers the installation of a fireplace as well as detailed information on the installation of a prefab chimney. Illus. 104 shows sections going through corner of kitchen. Illus. 103 shows fitting required for ceiling joists.

103

104

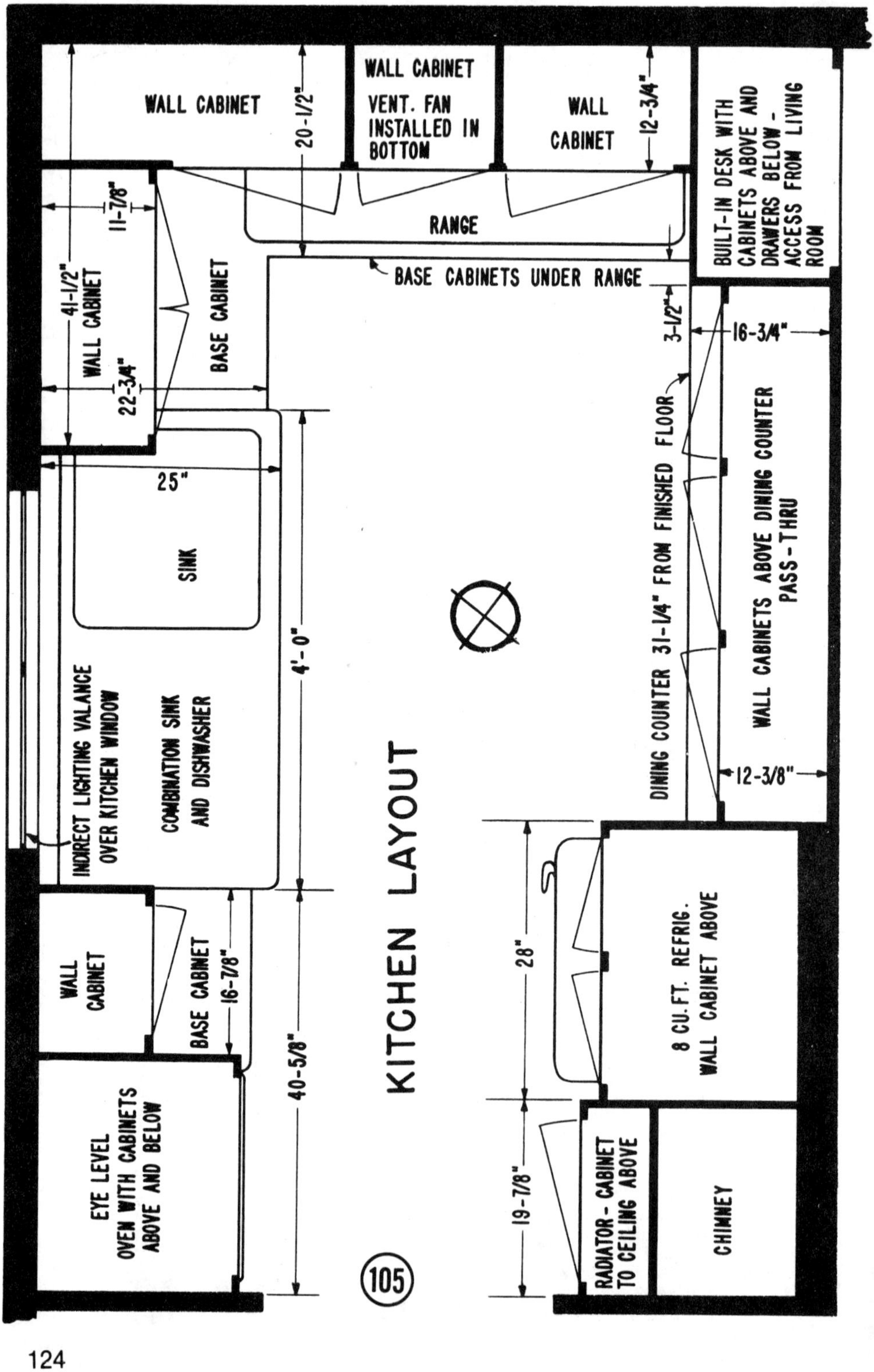
KITCHEN LAYOUT
105
WALL CABINET
WALL CABINET
VENT. FAN INSTALLED IN BOTTOM
WALL CABINET
20-1/2"
12-3/4"
BUILT-IN DESK WITH CABINETS ABOVE AND DRAWERS BELOW - ACCESS FROM LIVING ROOM
RANGE
BASE CABINETS UNDER RANGE
11-7/8"
41-1/2"
WALL CABINET
22-3/4"
BASE CABINET
3-1/2"
16-3/4"
25"
SINK
DINING COUNTER 31-1/4" FROM FINISHED FLOOR
WALL CABINETS ABOVE DINING COUNTER PASS-THRU
12-3/8"
4'-0"
INDIRECT LIGHTING VALANCE OVER KITCHEN WINDOW
COMBINATION SINK AND DISHWASHER
WALL CABINET
BASE CABINET
16-7/8"
28"
8 CU. FT. REFRIG. WALL CABINET ABOVE
EYE LEVEL OVEN WITH CABINETS ABOVE AND BELOW
40-5/8"
19-7/8"
RADIATOR - CABINET TO CEILING ABOVE
CHIMNEY

Check local building codes to make certain chimney meets requirements for heating system selected. Also check height of chimney above ridge required by local ordinance. Most prefabricated chimneys come in one and two foot sections, Illus. 104. They are easy to install and perform perfectly.

ROUGH-IN PLUMBING, WIRING, HEATING

At this stage all plumbing, wiring and heating should be roughed-in. Your heating contractor can recommend amount of radiation required to meet local conditions. Most telephone companies will provide underground cable between house and service pole. Your electrician can install necessary conduit to conceal telephone wire. He can also install television aerial lead-ins with outlet boxes where desired. Give consideration to adequate outside flood-lighting with switches in a bedroom as well as alongside front and back door.

In a test house, a TV was installed in a recessed opening below stairs in basement, also one in living room. Ventilation in recess was provided by lining back with a hardboard grille.

LIGHTING

To provide protection at night, install floodlights on gable ends of house. Consider running underground cable to floodlights in a tree or pole to illuminate front and rear of house. Position lights high to point down. These, coupled with front and rear porch lights, and a post light near driveway, should be controlled by a master switch placed next to the bed, with alternate switches in strategic locations.

If you plan on installing a burglar or fire alarm system, now is the time to rough-in wiring. Book #695 How to Install Protective Alarm Devices explains everything you need to know to install an alarm system.

Run necessary size wire to the laundry room to accommodate a combination washer and dryer.

Install wiring capable of servicing a dishwasher sink, counter range and wall oven, an air conditioner plus other appliances you may want. Discuss your needs with the utility company so you can obtain sufficient housepower. The added cost of installing proper size cable now is peanuts compared to what it will cost after the house is completed.

Recessed ceiling lights should be installed in kitchen, hall and recreation room, Illus. 3, 105. An outlet for a clock in hall, opposite kitchen, will prove convenient.

A fluorescent light fixture in bathroom, Illus. 3, can be concealed by a valance board. Install pull chain fixtures in all closets, utility room and in attic over stairs. Suggested wiring for downstairs rooms is indicated in Illus. 2.

BATHROOM

There are now many labor saving ways to rough-in plumbing for a bathroom, Illus. 3. It can be done in the conventional, and time consuming way with hub and spigot cast iron, Illus. 106; or in almost half the time with no-hub cast iron, plastic or copper, Illus. 107. Book #682 How to Add an Extra Bathroom provides all the basic information needed.

Those who want to rough-in copper or plastic supply and drainage lines find the new products easy to install. Preassembly of a plumbing wall using copper or plastic isn't difficult or same can be purchased preassembled, Illus. 108.

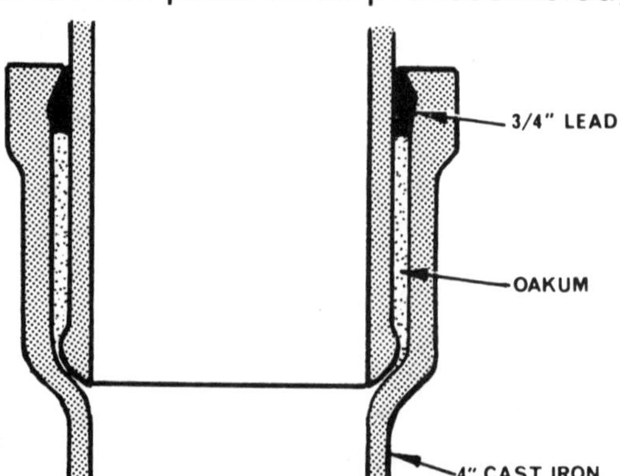

106

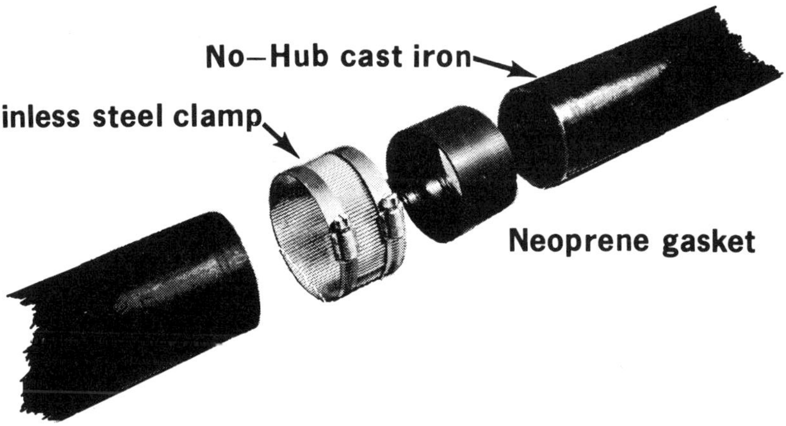
No–Hub cast iron
Stainless steel clamp
Neoprene gasket

107

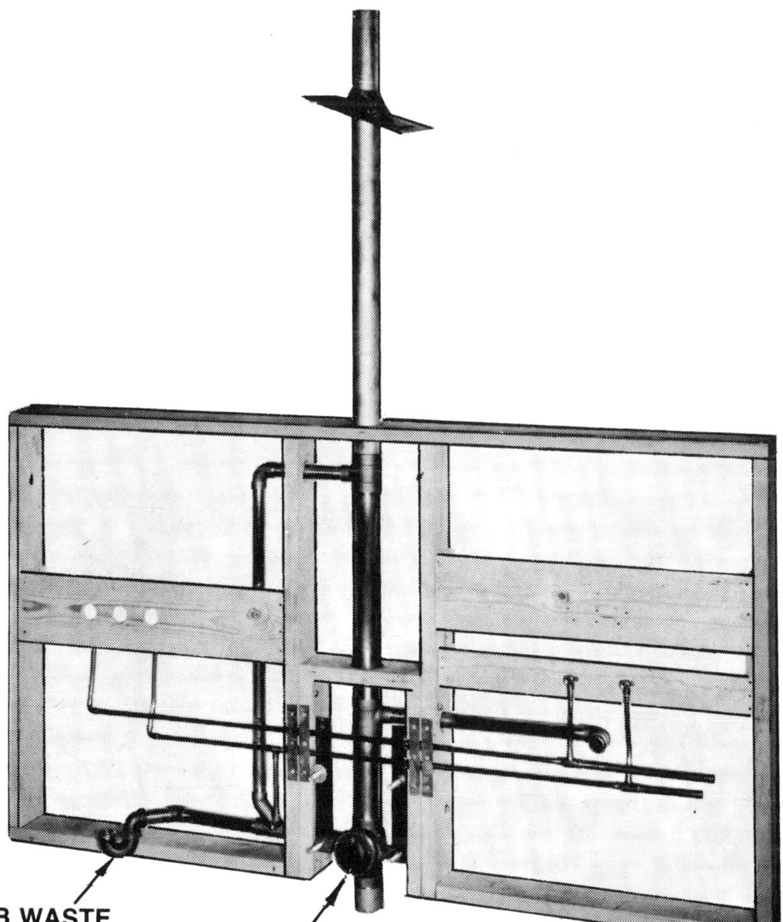
TUB WASTE
TOILET WASTE

108

Even more on the job time can be saved by using a copper or plastic pre-engineered plumbing wall, Illus. 108. These are available with a wall hung closet, tub and lavatory, Illus. 109.

Also available is a completely prefabricated bathroom, Illus. 110. This comes in four sections, Illus. 111.

Illus. 6 shows how a pre-engineered plumbing wall can be installed to service both a bathroom and kitchen. This installation requires 2 x 6 framing. The plumbing wall is shipped completely assembled. This necessitates using 2 x 6 shoe, plate and studs. The assembled unit requires 88" of wall space.

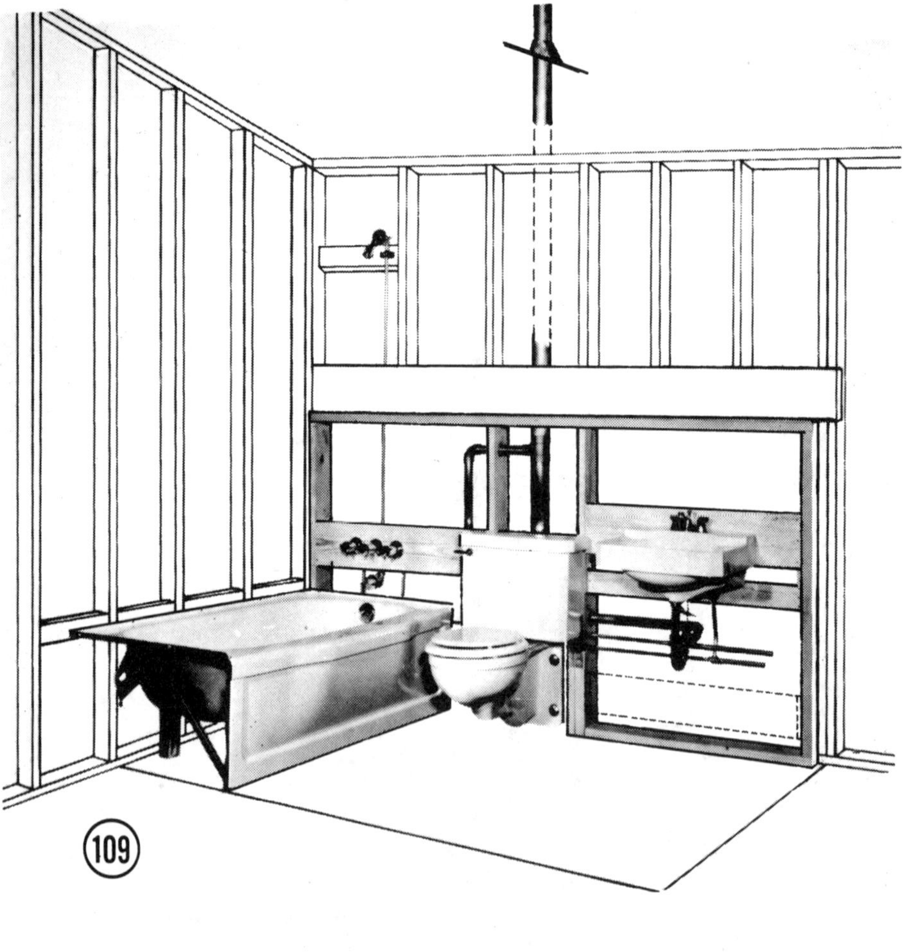

109

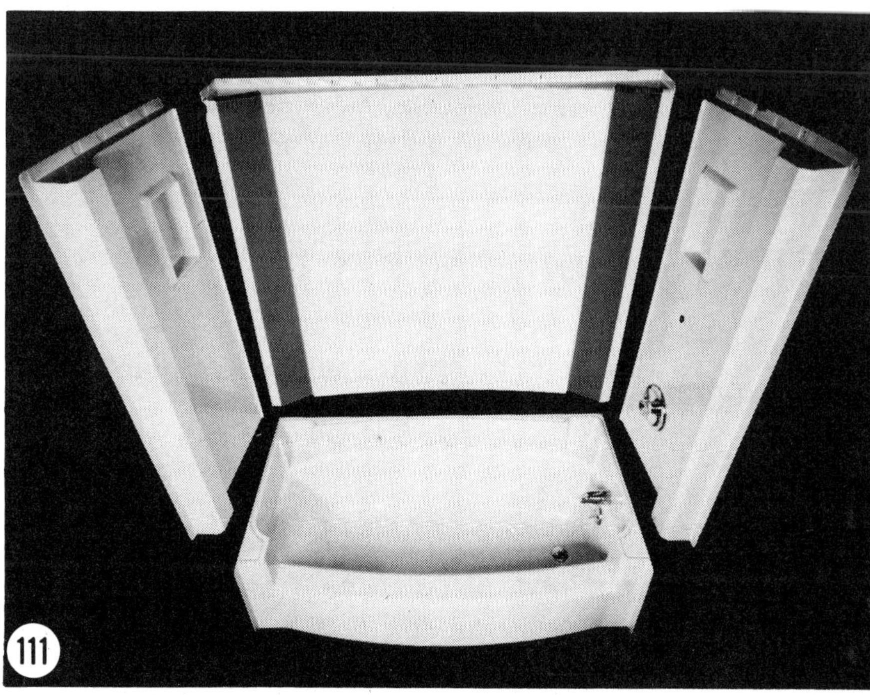

APPLY CLAPBOARD

¾ x 8" siding to cover 170 sq.ft.

If you use wood siding, prime coat clapboard on both sides. Apply clapboard to gable ends with 6 penny aluminum nails. Start at drip cap on gable ends and work up, Illus. 13,94. Bevel bottom of first piece of clapboard to butt tightly against flashing over drip cap. Cut ends to angle required so clapboard butts tightly against gable fascia. Expose about 7" of clapboard on each course. To apply clapboard like a pro, lay out each course and see if the top course butts neatly under lip on louver. It's permissible to allow 7¼, even 7½" exposure on some courses if necessary.

EXTERIOR SIDING AND PORCH CEILING

Approx. 27 panels — ⅜" x 4 x 8' exterior grade plywood.

You can apply clapboard siding to entire house or use clapboard and exterior grade plywood where indicated, Illus. 11,12,13,14. Prime coat plywood before applying. Nail siding with 6 penny aluminum nails every 6" on studs. Butt panels together over studs. Check first panel with level to make certain it's plumb. Butt panels tight against corner boards, fascia boards and window trim. Putty all nail holes and cracks. Steel wool prime coat, then apply two coats of outside paint.

OPTIONAL SIDING EFFECT

You can apply ⅜" white pine mullion every 4' to conceal joints, or every 16" if you want to achieve a colonial batten effect, Illus. 112,113.

PORCH CEILING

There are now many different exterior grade panel boards available that are suited for porch ceilings. If you prefer to use one of these instead of plywood, buy quantity required. Follow manufacturer's directions when installing.

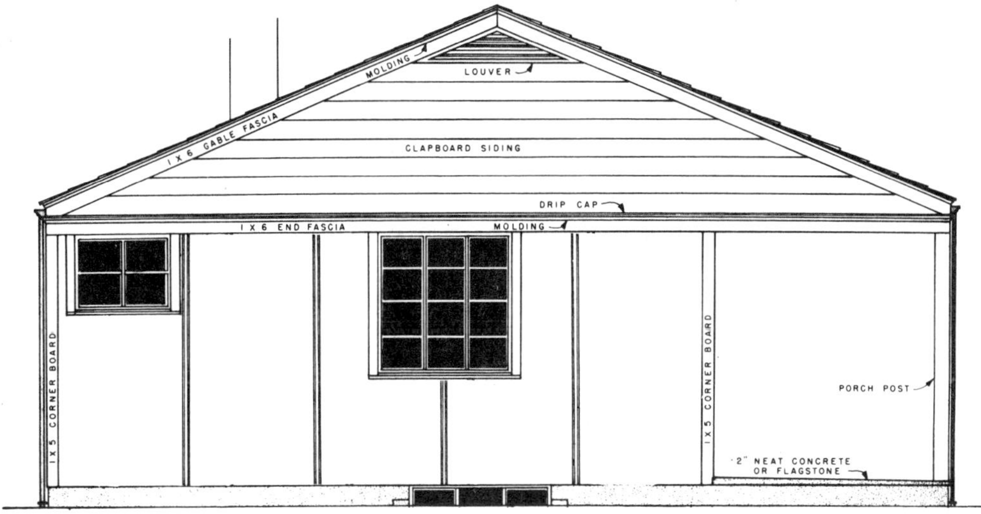
MOLDING
LOUVER
1 X 6 GABLE FASCIA
CLAPBOARD SIDING
DRIP CAP
1 X 6 END FASCIA
MOLDING
1 X 5 CORNER BOARD
1 X 5 CORNER BOARD
PORCH POST
2" NEAT CONCRETE
OR FLAGSTONE

112

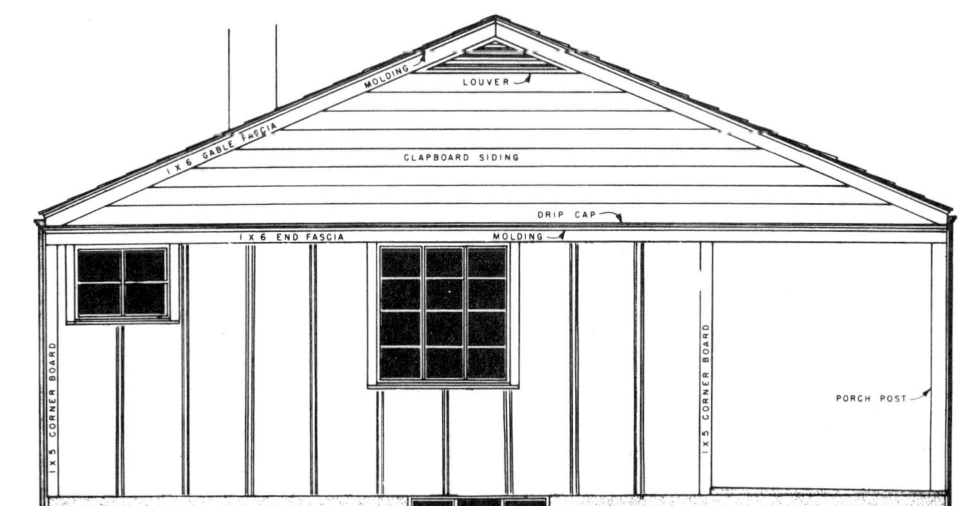
MOLDING
LOUVER
1 X 6 GABLE FASCIA
CLAPBOARD SIDING
DRIP CAP
1 X 6 END FASCIA
MOLDING
1 X 5 CORNER BOARD
1 X 5 CORNER BOARD
PORCH POST

113

HANG EXTERIOR DOORS

Front — 1 - 2'8" x 6'8" Dutch or other exterior door
Rear — 1 - 2'8" x 6'8" Exterior door

Many lumber and hardware stores rent 6' levels. These come in handy when setting door frames.

Buy exterior doors with assembled frames to size rough opening requires.

Since the casing on some outside door frames is nailed directly to studs, while others are nailed over sheathing, Illus. 115, tell your dealer what thickness sub-sheathing, and what thickness interior wallboard you plan on using. He will then be able to advise you concerning installation.

You will need four 4 x 4" loose pin brass butt hinges to hang Dutch doors, Illus. 114; three 3½ x 3½" loose pin brass hinges to hang an exterior door.

There are several important facets to installing an outside door frame:

1. The top inside edge of door sill should be flush with finished floor, Illus. 116.
2. The door sill must be level, the door frame plumb and square.
3. The width of door jamb, Illus. 115, permits nailing outside casing either against studding, or over sub-sheathing, while the inside casing nails directly over interior paneling or gypsum board.
4. Use a filler strip, Illus. 115, when you can't purchase a frame cut to size required. Use filler block on inside as shown.

114

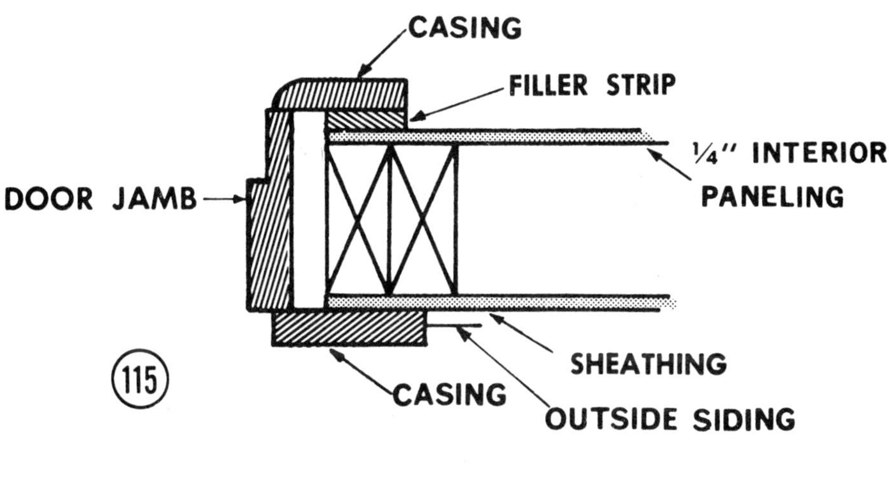
CASING
FILLER STRIP
1/4" INTERIOR
PANELING
DOOR JAMB
SHEATHING
CASING
OUTSIDE SIDING

115

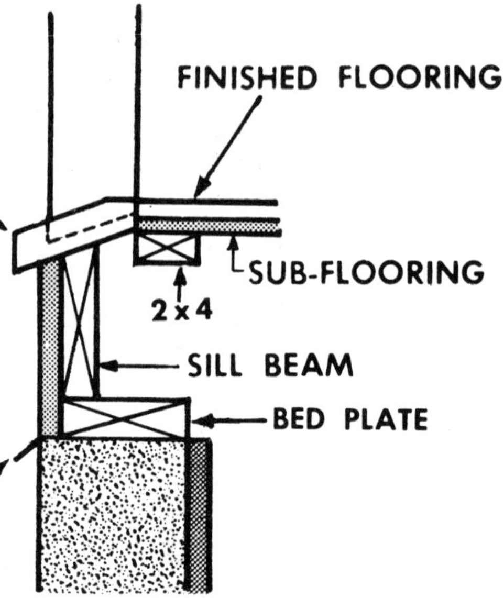

116

Saw shoe off flush with studs at door opening, Illus. 117.

Saw subfloring to width and depth door sill requires. Saw sill beam down 1½ to 2" depth and angle required.

To stiffen edge of subflooring, nail a 2 x 4 in position shown, Illus. 116. Allow 2 x 4 to extend 6 to 8" beyond edge of opening, then toenail 2 x 4 to floor. Drive 8 penny nails through floor into 2 x 4. Or you can install solid bridging, Illus. 61, and drive nails through bridging into ends of 2 x 4.

Your first objective in setting a door frame is to position it in opening so edge of door sill is level with finished flooring, Illus. 116, you have selected. Tack a piece of finished flooring, tile, underlayment or carpet, etc., in position against sill to indicate height of finished flooring.

Place level on sill and tack frame temporarily in level position.

To make certain door frame is plumb, do this. With square, locate and draw line at center of header and also on sill. Drive a 6 penny finishing nail temporarily in front center of header.

Tie a plumb bob line to nail. When point of bob is directly over center line on sill, frame is considered plumb. Or use a 6' level to plumb frame.

When frame is level and plumb, nail in position with 8 penny finishing nails through casing.

Don't drive nails all the way. Check corners of frame with carpenter's square. When square, level and plumb, drive 8 penny finishing nails through casing into studs every 12". Countersink heads, fill holes with putty. Remove temporary bracing.

Nail filler blocks A, Illus. 117, in position indicated opposite hinges and strike plate. Use thickness plywood space requires, but don't force jamb out of position. These blocks stiffen framing opposite hinges and strike plate on door lock.

Pack loose rock wool insulation between frame and stud.

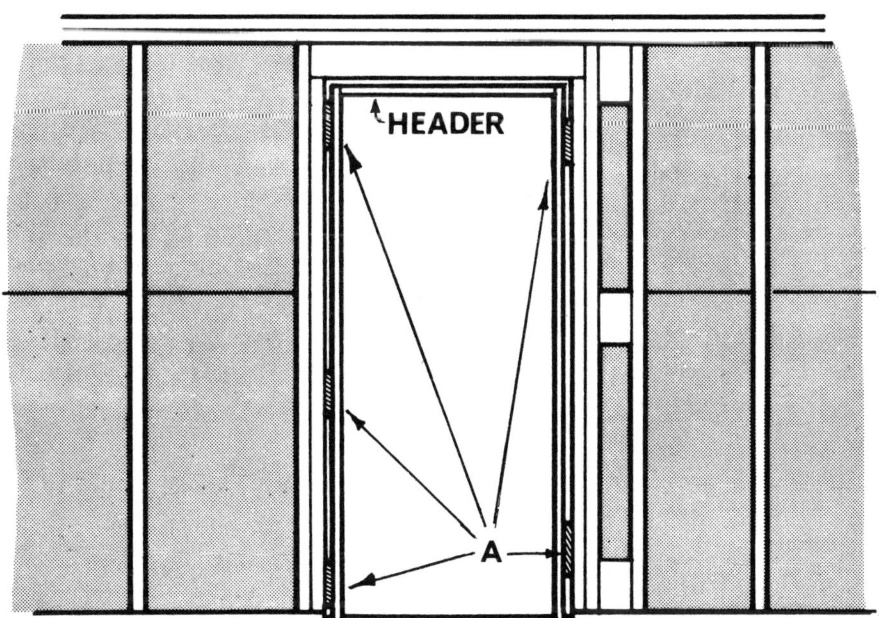

117

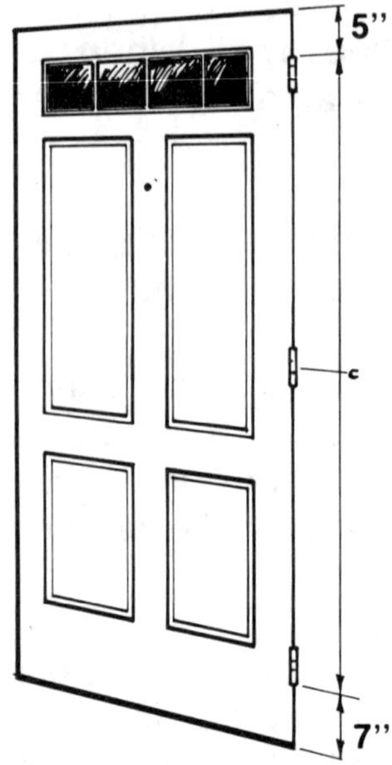

118

Most pros place hinges for a 2'8" outside door in position indicated, Illus. 118. This is usually 5" to top hinge; 7" up from bottom to lower hinge. When a third hinge is used, it is centered between the two.

Larger doors are usually hinged 7" down to top of upper hinge; 9" up to bottom of lower hinge.

Hinges on Dutch doors, Illus. 114, are usually placed in position indicated to line up with panels. The bottom edge of lower hinge lines up with panel; the top edge of upper hinge also is in line with panel.

3½ x 3½" or 4 x 4" loose pin brass hinges are recommended. Take hinge apart, place in position and mark outline on door, Illus. 119.

Use a 1" chisel, make cuts to depth equal to thickness of hinge leaf, Illus. 120.

Chisel out mortise, Illus. 121, and fasten leaf in position with screws provided.

Follow same procedure and mortise out jamb, Illus. 122.

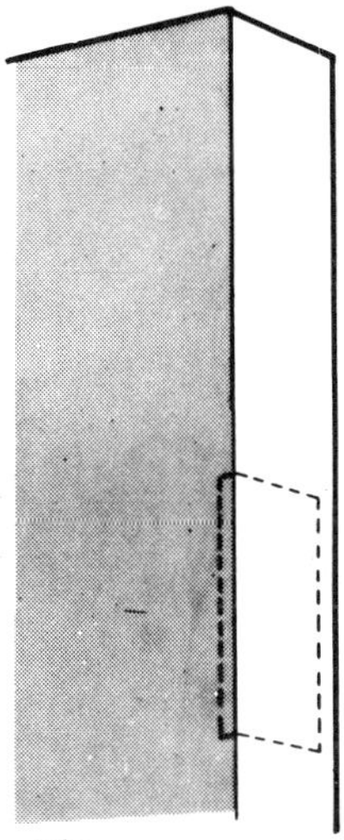

119

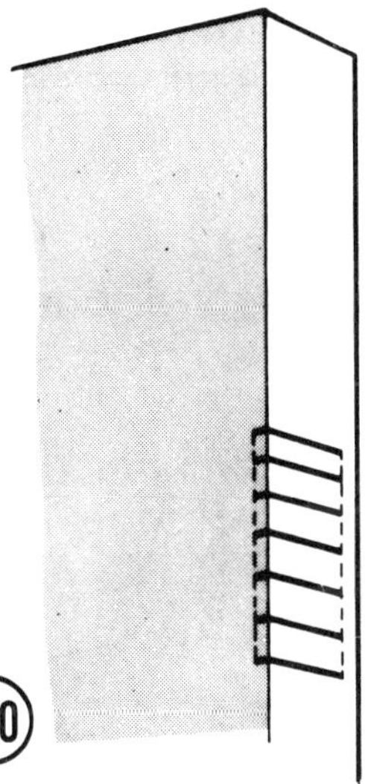

120

121

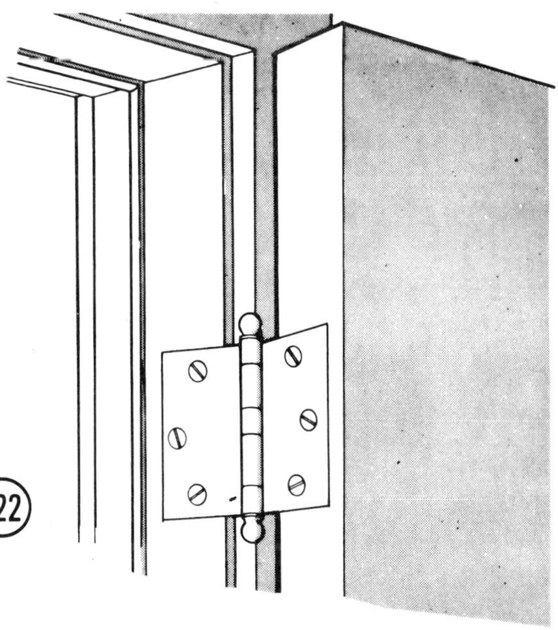

122

BASEMENT FLOOR

2 x 3 — 290 lineal feet for sleepers (A).

CAUTION: Since a basement is vulnerable to a heavy rain, a stopped up gutter, etc.; make provision for a sump pump. Books #615 How to Modernize a Basement and #675 Plumbing Repairs Simplified provide complete details.

When the plumber, electrician and heating contractor have completed roughing-in, lay concrete floor in basement.

Spread crushed stone over floor area level with partition footings E, Illus. 123,124. Pour a 2" layer of concrete B. Mix waterproofing in concrete for subfloor B and top layer C.

To permit anchoring finished wood flooring or underlayment for carpet, use 2 x 3 sleepers A. Paint these with wood preservative. Drive 8" nails every 16" into edges, allowing nails to project at least 1 to 1¼".

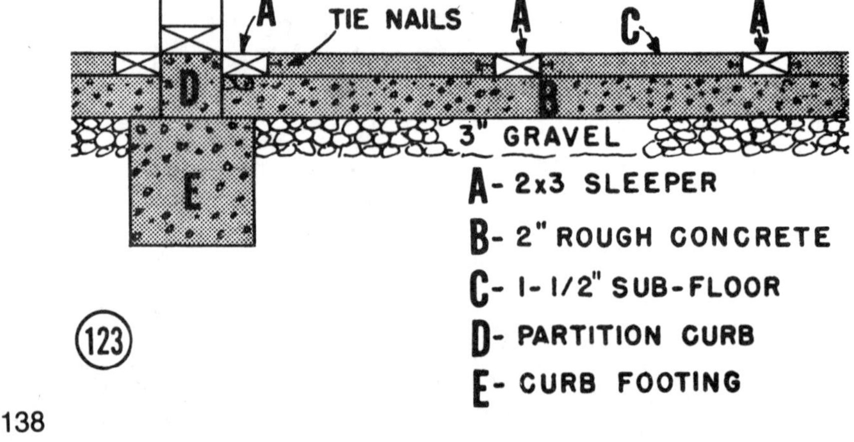

123

Space sleepers A, 16" apart. Level A with top of D. Install sleepers on opposite walls. Level these and stretch a line across to insure laying others at height required. Pour 1½" concrete C, level with sleepers, Illus. 123,124. Allow concrete to set at least three days.

Before laying finished wood flooring or underlayment for carpeting, lay a moisture barrier. You can either paint floor area with asphalt roofing, then embed #15 felt. Or lay one piece of polyethylene over entire area.Tack sides about 3" up walls. It can be cut off flush with finished flooring or after running baseboard.

If you want to lay ceramic or patio tile in basement, do not use sleepers or moisture barrier on top course of concrete. Read Book #816 How to Lay Ceramic Tile. Also consult Book #617 Concrete Work Simplified for additional information on concrete work.

FACTS ABOUT STAIR CONSTRUCTION

1 x 12 — 2/12
2 x 10 — 2/10
2 x 8 — 2/10
2 x 4 — 1/6, 1/10
1 x 2 — 2/10
5/8'' plywood — 4 x 4'

Since most lumber dealers sell prefabricated stairways, we recommend buying, rather than building. If you decide to build, use materials suggested above. If you buy, ask dealer to check your job to obtain exact dimensions required.

If you prefer building, build platform to size indicated, Illus. 125. Cut 2 x 4 legs to length required to position 5/8'' plywood platform 6'0'' minimum from bottom of 2 x 8 floor joists. (If you have increased basement ceiling height one additional course of block, as mentioned on page 43, height from platform to joists will be approximately 6'8''.

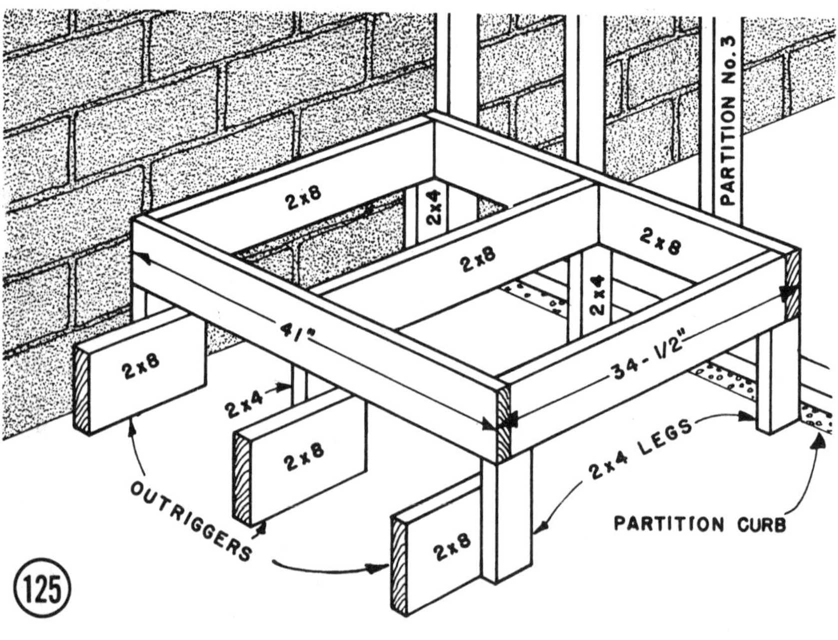

125

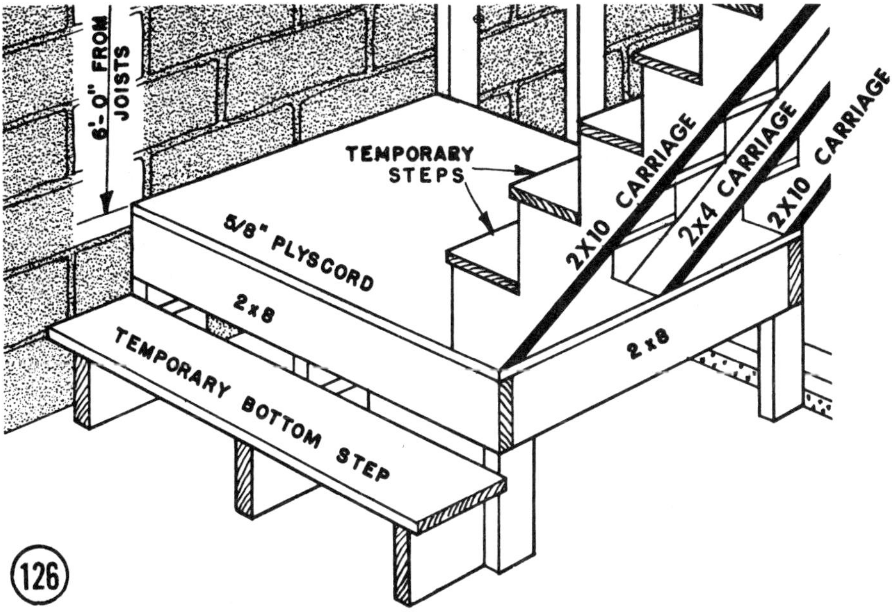

(126)

Use 2 x 8 for frame. Nail legs to frame. Reinforce legs at corner with a scrap length of 2 x 4 or 2 x 8, Illus. 126.

Bottom step rests on 2 x 8 outriggers nailed to legs of platform. Use 5/8" plywood for platform. For temporary treads, use sheathing or 5/8 or 3/4" plywood.

1 x 12 stair stringer is cut to length and same top and bottom angle as carriage. After cutting carriage to shape and length required, nail 1 x 2 in position shown. Place stringer on top of 1 x 2 and nail stringer to carriage.

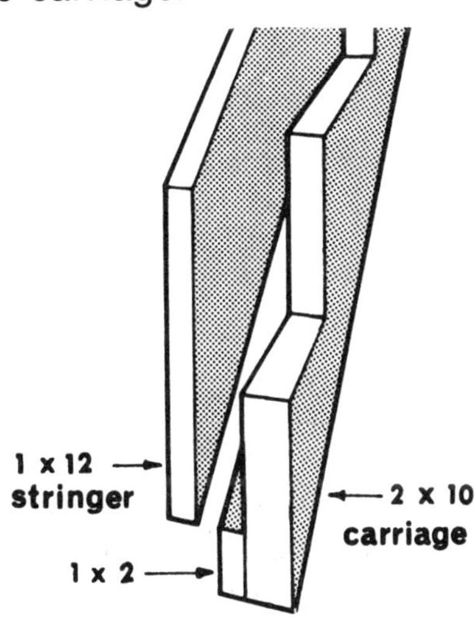

(127)

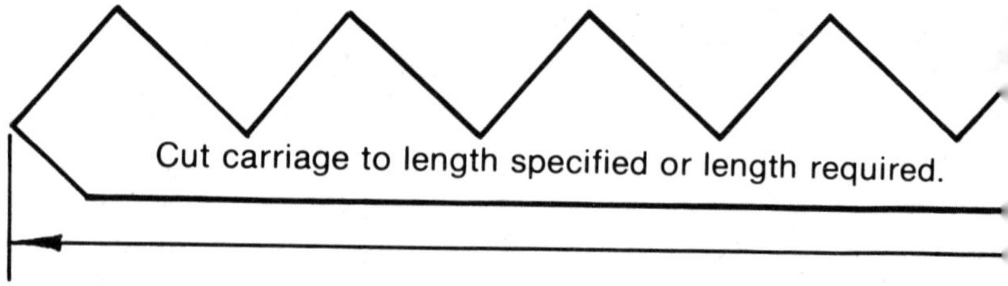

128

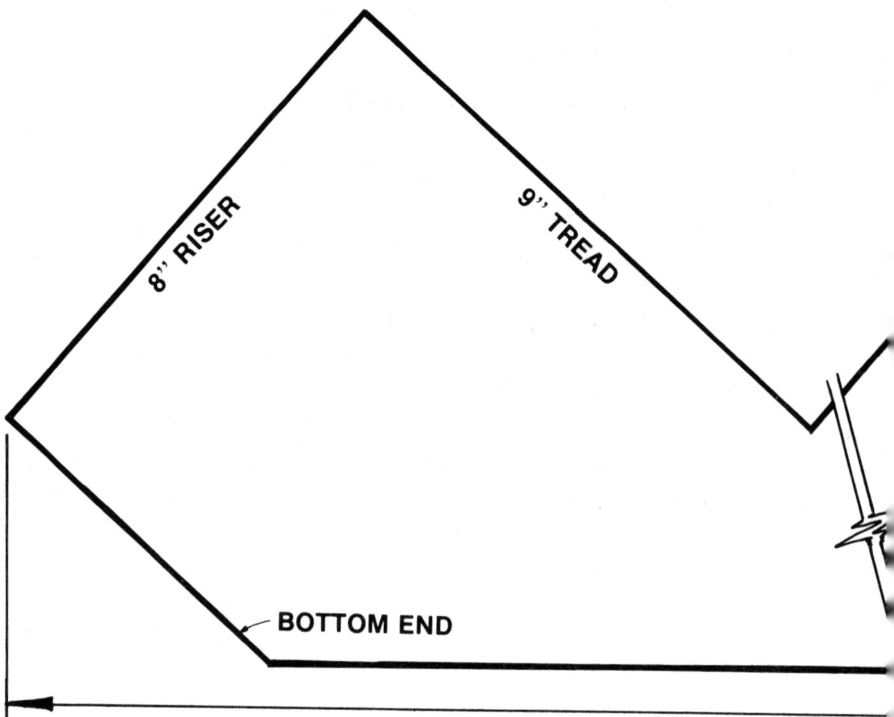

129

Use 1 x 12 for stringers, 2 x 10 for carriages. Cut a 2 x 4 x 10' to angle at bottom and to length and angle at top shown in Illus. 128, 129, or to length required. Use this to test length of carriage required for your installation. If it fits, cut two 2 x 10 carriages to angle shown, Illus. 129.

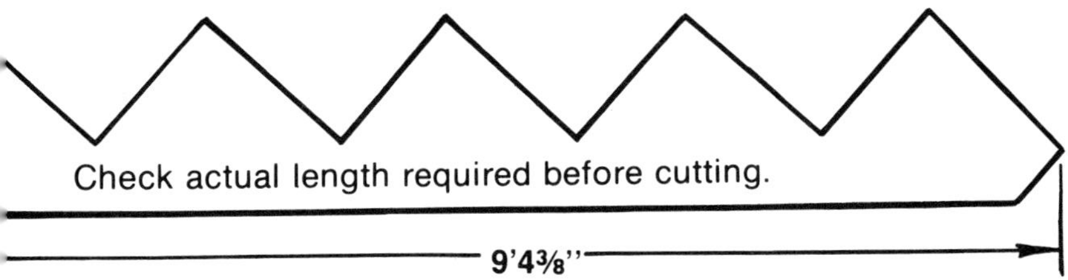

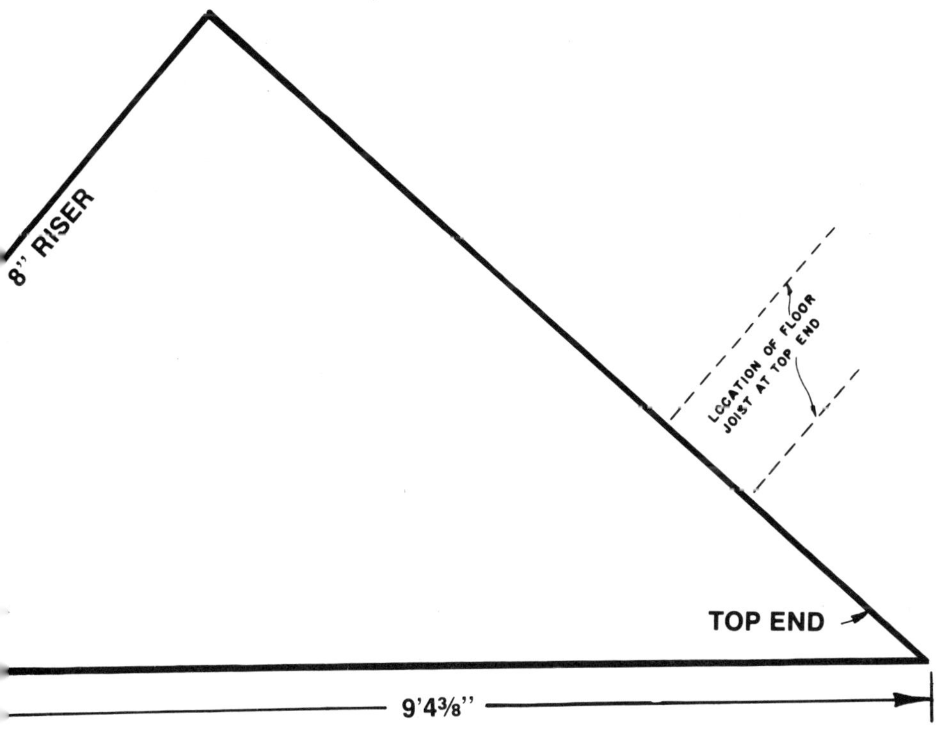

If you buy prefabricated stairs that require assembly on the job, ask your dealer to measure job and provide one man when same is installed. If you act as a helper, you can save a considerable amount.

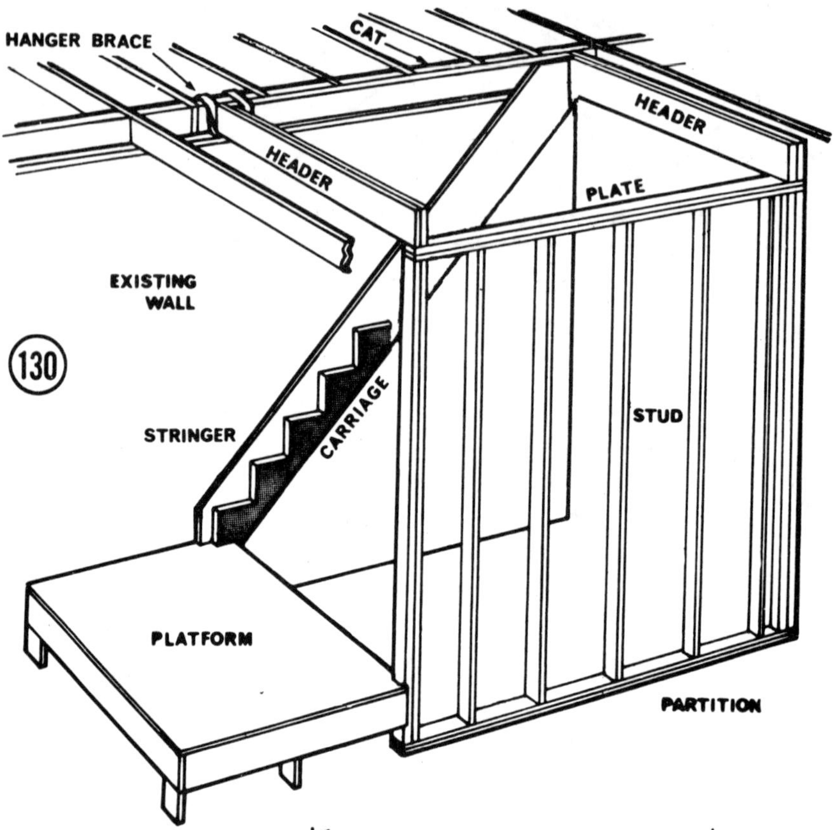
HANGER BRACE
CAT
HEADER
HEADER
PLATE
EXISTING
WALL
STRINGER
CARRIAGE
STUD
PLATFORM
PARTITION

(130)

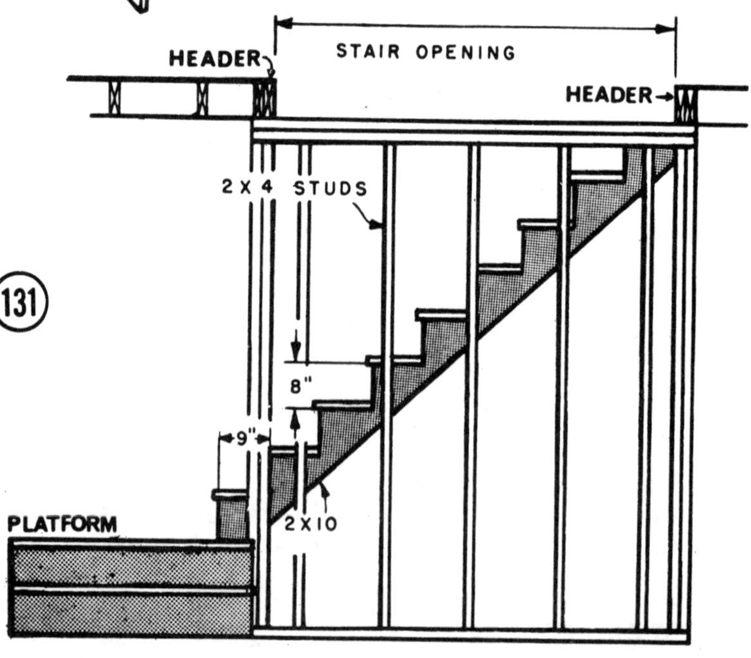
HEADER
STAIR OPENING
HEADER
2 X 4 STUDS
8"
9"
2 X 10
PLATFORM

(131)

We show 8'' risers, 9'' tread, Illus. 130,131. Most millwork houses sell oak treads up to 12'' wide and will cut these down to size required. Risers up to 9'' are also available. If you use a 9'' riser, use a wide (11½'') tread. You can alter height of riser from 7¼ to 8'' if you find it necessary to compensate for any variation in construction. Buy precut treads and risers.

Stringers should finish flush with finished flooring. Nail 1 x 2 to 2 x 10 carriage in position shown, Illus. 127.

When laying out stairs, select height of riser required.

Nail stringer to carriage, then nail assembled stringer and carriage to platform, header and studs, Illus. 130. Nail cutouts from 2 x 10 carriage to 2 x 4, Illus. 126, and use this for a middle carriage. This provides more support for stairs, also a nailor for gypsum board on closet ceiling.

Apply glue to each riser and nail it in place with 8 penny finishing nails. Apply glue to bottom of tread where it butts against carriage, and to back edge where it butts against riser.

Draw lines on each tread to indicate exact center of riser. Drill holes through tread and nail through tread into riser, Illus. 132, with 10 penny finishing nails.

Nail through riser into back edge of tread with 6 penny box nails, Illus. 133.

We recommend nailing ¾'' plywood to carriage for temporary steps during construction, Illus. 126. If finished treads and risers are installed, cover same with building paper until job is completed.

The stringers on most prefabricated stairs are routed out to accommodate tread and risers. Wedges, supplied by stair builder, are glued and driven in position under treads and risers to lock both in position.

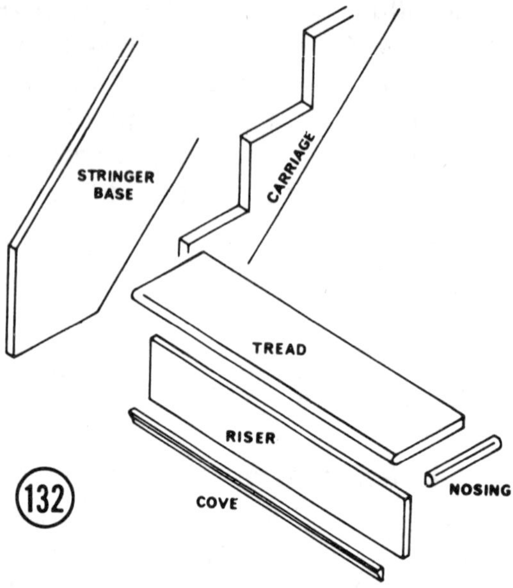

132

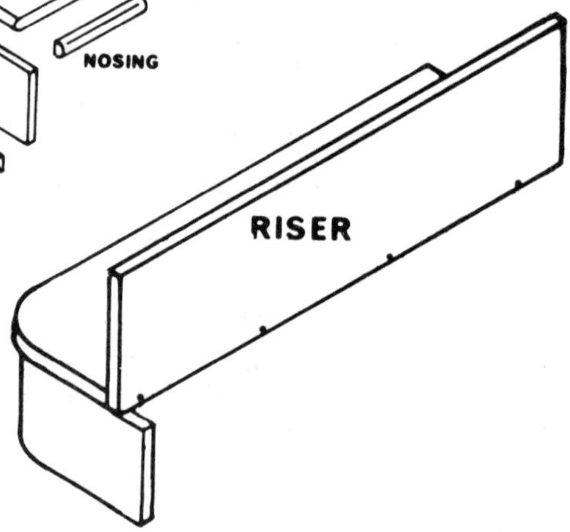

133

The carriage is notched with size and number of tread and risers height requires. If ceiling height or floor space requires a platform, Illus. 131, same can be used.

Most building material suppliers sell stock treads and risers. Let's say you buy 8" risers and 10" treads. When you nail the first two risers in position, and then the first tread, it will project over face of lower riser approximately 1". When you nail cove molding, Illus. 132, in position, it trims out stairs.

If you install wrought iron stair rails, use brackets and directions manufacturer provides.

BUILD PARTITIONS

2 x 4 — 1/16,2/12,1/10
2 x 6 — 1/6

Build partitions 4 and 5 in position indicated, Illus. 45, to overall size indicated in Illus. 49,50, or to size required.

FURRING - PANELING

Use 5/4 x 3 or 1 x 4.

To eliminate dampness, paint inside of foundation walls with two coats of waterproofing following directions manufacturer recommends.

Apply furring to walls of recreation and laundry rooms. 8 penny steel cut nails can be driven into mortar joints, Illus. 134.

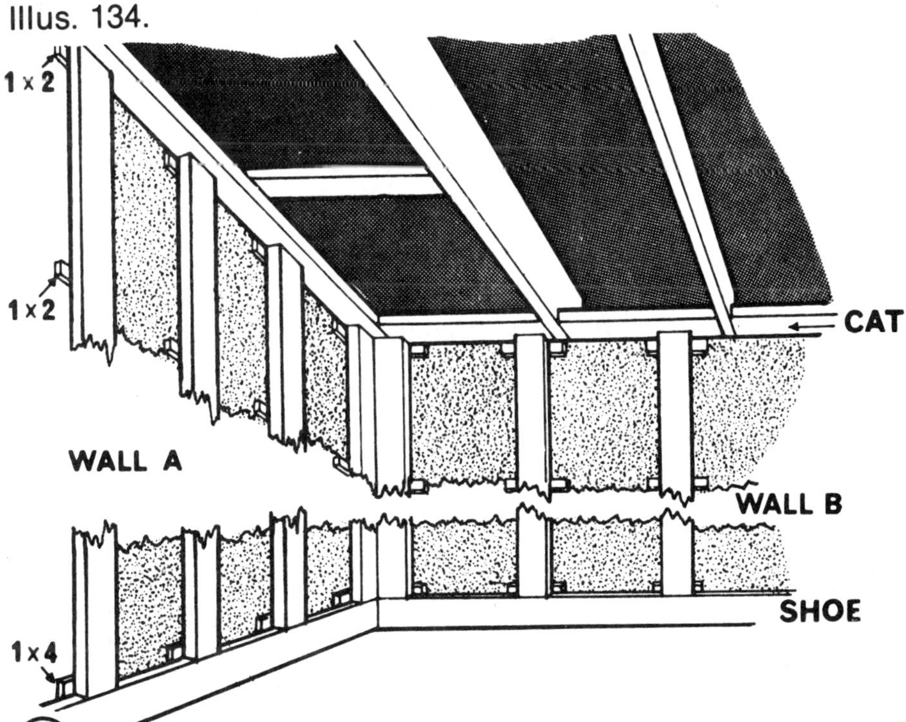

(134)

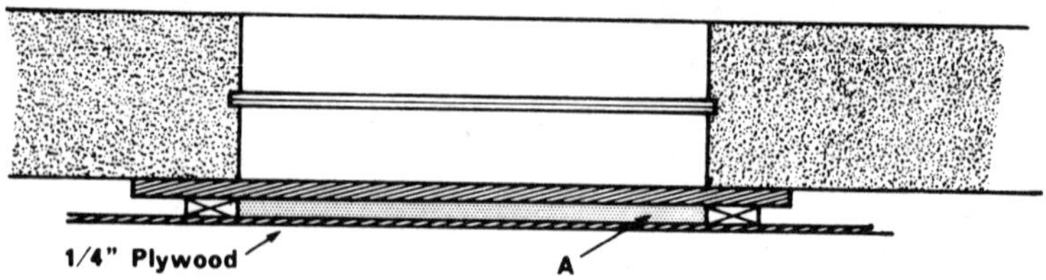

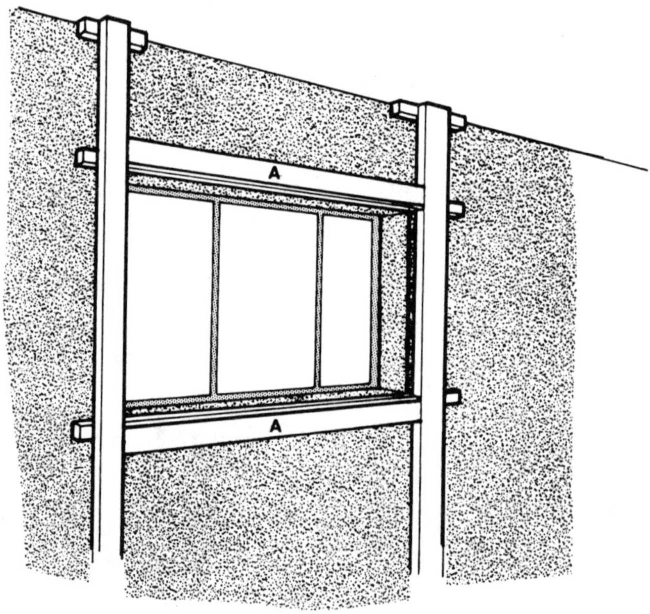

135

Apply furring A around basement windows as shown, Illus. 135,

When applying panels, cut each so it projects ¼" beyond edge of window recess all the way around. Next cut plywood to size required for sides of recess. Nail these in position to edge of furring. Cut plywood to size ceiling of recess requires so ceiling wedges against sides. Nail ceiling to furring strip. Cut piece of plywood for bottom so it wedges against sides. Nail bottom to furring. Apply Putty Stik to exposed edge of plywood and you finish trimming window like a pro. Casing, Illus. 136, can be used to trim out window. Miter cut ends.

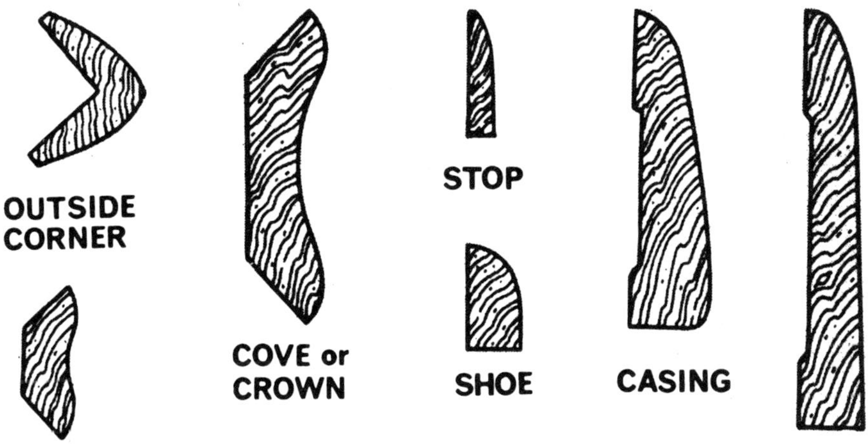

DESCRIPTION	SIZE	LTH.
Shoe	5/8" x 5/16"	10′
Stop	1 5/16" x 5/16"	10′
Casing	1/2" x 2 1/4"	7′
Casing	1/2" x 2 1/4"	10′
Base	13/32" x 3"	10′
Outside Corner	7/8" x 7/8"	8′
Inside Corner	3/4" x 3/4"	8′
Cove	9/16" x 2 1/4"	10′

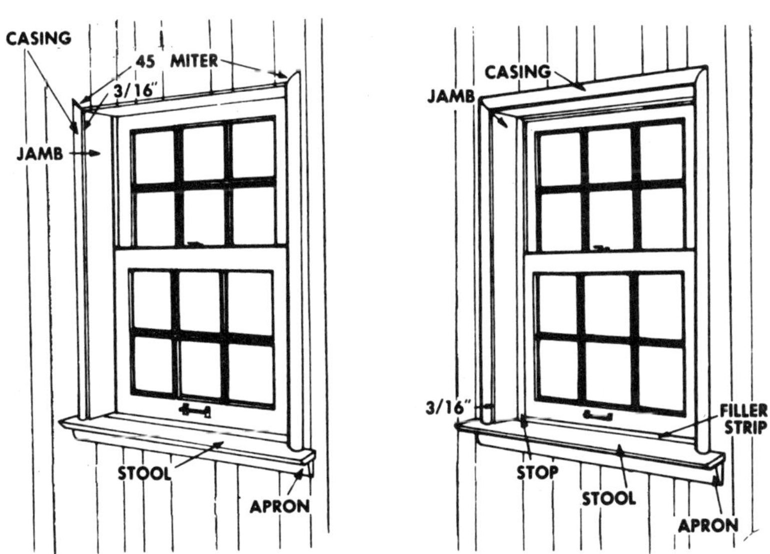

136

INSULATION

Staple aluminum foil backed insulation or rigid fiberglass panels between furring before applying paneling to walls.

Staple insulation between studs on all outside wall frames and between ceiling joists or rafters. If house is built over a crawl space, staple insulation between floor joists. To make certain it stays up, nail wire netting (chicken wire) across joists. Follow manufacturer's directions for insulation selected.

INSTALL CEILINGS

Gypsum board is recommended. It is easy to install and provides years of service. To simplify handling the 4 x 8' panels, make T supports using a 1 x 6 x 4' nailed to the top of a 2 x 4 cut to length ceiling height requires.

Complete details concerning installation of a luminous ceiling is provided in Book #773 How to Create Room at the Top.

LIVING ROOM WALLS

Approx. 9 — ¼" x 4 x 8' panels will be required.

Hardwood plywood or wood grained hardboard or gypsum board provides a quick, easy way to finish a room.

To estimate exact number of panels a wall requires, measure overall length, including doors and windows and divide by four. Place panels in room where they are to be installed to allow them to become acclimated. To match grain and obtain a pleasing selection of color, place panels in position around room.

Cut panels ⅛ to ¼" less than overall height required.

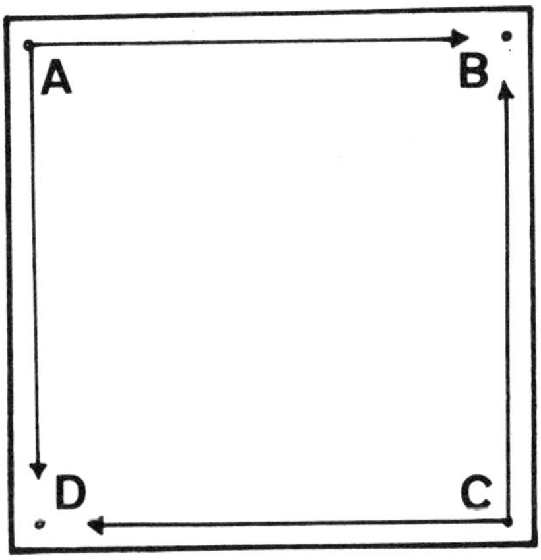

137

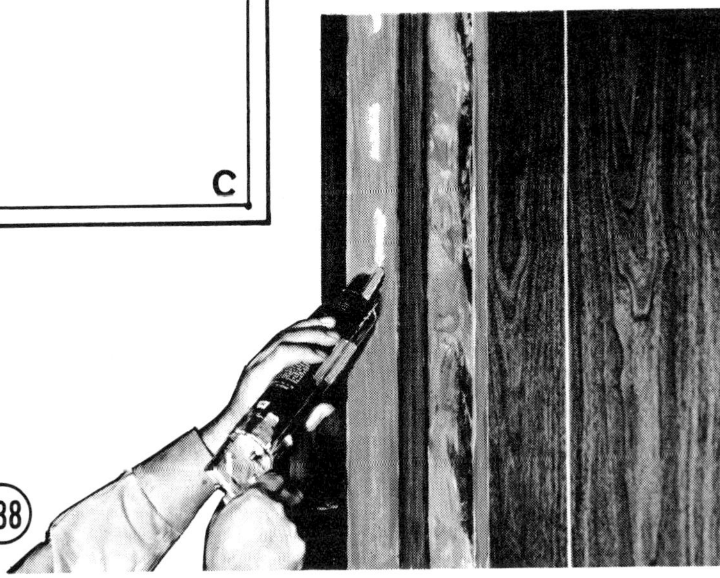

138

Always start in corner A, Illus. 137, and work toward B, cutting last panel to fit. Start at C and work toward B. Start at C and work toward D. Start at A and work toward D.

Place first panel in position and check with level. If plumb, apply 3'' strips of panel adhesive every 6'' to each stud, Illus. 138. Some manufacturers suggest running a continuous bead of adhesive along shoe, plate, cats, header and on those studs that back edge of panel. Never jam edges of panels together. Press panel in position following adhesive manufacturer's directions.

Always place first panel in position without adhesive to make certain it's plumb. If it isn't plumb, scribe it to corner, Illus. 139. When the first panel is plumb, the others line up properly.

Panels can also be applied with matching color nails.

139 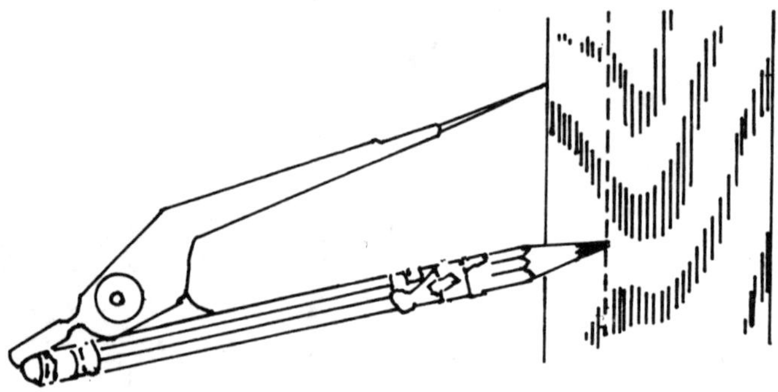

There are many simplified steps to paneling that are described in detail in Book #605 How to Install Paneling.

The walls of the living room will require approximately 288 square feet of ¼" hardwood plywood or gypsum board; the downstairs recreation room, 550 square feet; the bedroom and laundry room 650 square feet.

Many who build prefer installing plastic laminate on walls of a kitchen and bathroom. This provides a truly carefree wall covering. Nail ⅜" exterior grade plywood or vinyl covered gypsum board to studs before applying plastic laminate.

Ceramic and quarry tile floors are now being used in many more rooms than the bathroom. This is an ideal floor covering for kitchen, hall and downstairs rooms. Read Book #816 How to Lay Ceramic and Quarry Tile. It explains most of the "tricks of the trade."

Bathroom walls should be sheathed with vinyl covered gypsum board if you plan on installing ceramic tile. Only use vinyl covered gypsum board on areas to be tiled. Use regular gypsum board on any area that is to be painted or wallpapered.

Many building codes require the utility room or any room containing a furnace to be sheathed with asbestos cement board, aluminum or other metal. Asbestos cement board can be nailed directly to studs.

KITCHEN PLANNING

Base, wall and counter range cabinets, wall oven and refrigerator enclosure can be built following step-by-step directions described in Book #758 How to Modernize a Kitchen.

The counter in the "pass through," Illus. 140, makes a 36 x 51" table thanks to a fold down leaf in the living room side. ¾" matching plywood is fastened to counter with continuous hinge and table brackets. Knee space under counter in kitchen permits storing two stools.

A counter range can be built in where indicated, Illus. 105. A ventilating fan, installed directly overhead, draws cooking odors out of kitchen.

Follow manufacturer's directions when installing a flue and switch for ventilating fan.

Space allocated for sink will also accommodate a combination dishwasher-sink. Select a wall oven, range, refrigerator, sink and dishwasher that fit space available.

Install wall oven at eye level. Follow manufacturer's directions concerning ventilation and wiring. Construction of enclosure is described in Book #758.

Build all base cabinets to size space permits, then apply plastic laminate bonded to ¾" exterior grade plywood as a countertop. The retailer will specify size of opening required for sink and range you select.

As previously mentioned, nail ⅜" exterior grade plywood to studs below wall cabinets if you want to install a plastic laminate backsplash.

Since the kitchen is visible from the living room, use plywood on cabinets that matches plywood selected for living room.

FINISHED FLOORS

Oak flooring usually runs 13/16 x 2¼". You will need sufficient flooring to cover approximately 700 square feet. If you staple #15 roofing felt over subflooring, it deadens sound.

Nail finished flooring across floor joists in first floor rooms.

Nail finished flooring across sleepers in basement. Stagger all butt joints.

Since indoor/outdoor carpeting is now in popular use, give serious consideration to the many advantages it offers.

If carpeting is installed, follow step-by-step directions and install underlayment to thickness specified in Book #683 Carpeting Simplified.

HANG DOORS

Set interior door frames level, plumb and square following directions previously outlined. Shim frame to stud where hinges and strike plate are installed.

2 — 2'0" x 6'8" flush panel doors for living room closet and bathroom.

2 — 2'6" x 6'8" flush panel doors for bedroom, laundry and utility room.

5 prs. 7" "H" hinges.

5 matching latch sets.

While framing to kitchen permits installing a 2'4" door, we don't recommend hanging one. For those who want a bit more privacy in kitchen, hang 30 or 48" swinging cafe doors just below eye level height.

7" black colonial H hinges or 3½ x 3½" loose pin butt hinges can be used on all interior doors.

Frame opening to closet in bedroom for sliding or accordion folding doors. These come complete with track and hardware required. Book #605 How to Apply Paneling also explains how to build wall to wall closets.

A 14" wide plywood shelf and clothespole should be installed in closet. A sliding door can be installed in combination linen and housekeeping closet in bathroom.

APPLY TRIM

If baseboard heating is installed, only apply a baseboard in areas where there's no radiation. Install baseboard to same height.

Illus. 136 shows stock trim now available. This comes finished to match plywood selected. If gypsum board is installed and painted or covered with wallpaper, select unfinished trim and paint color desired.

Install door hardware following directions manufacturer provides.

Trim windows as shown, Illus. 141; interior doors, Illus. 142.

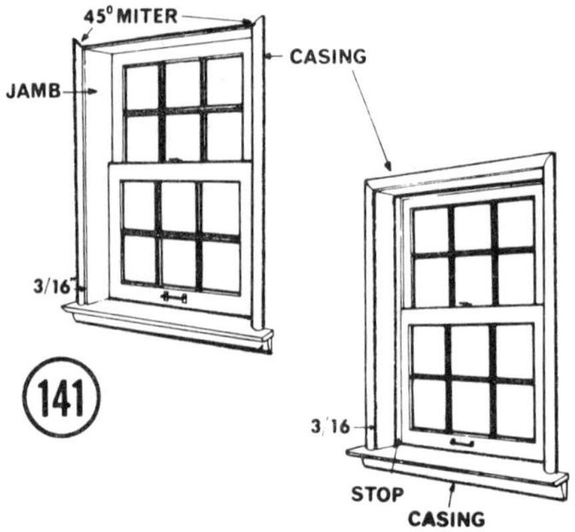

(141)

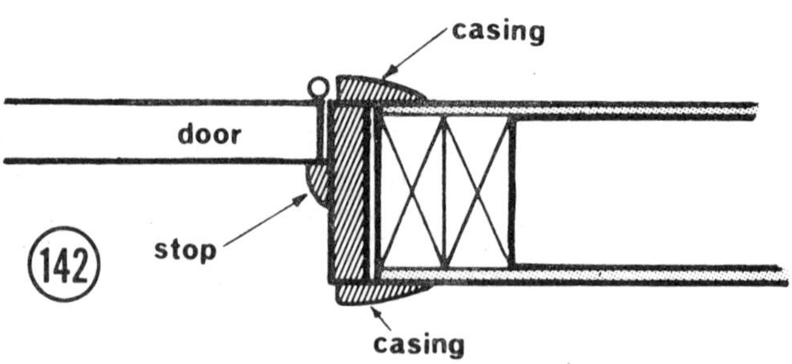

(142)

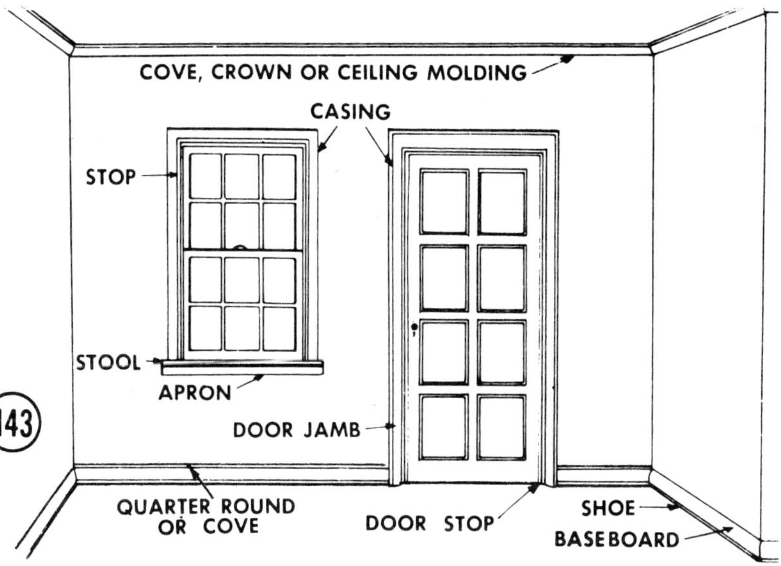

143

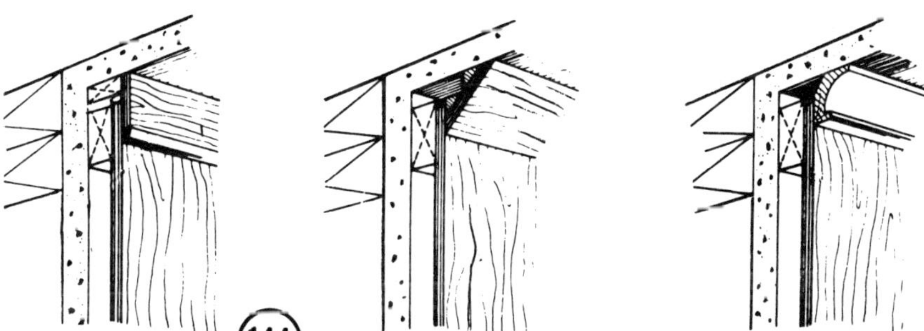

144

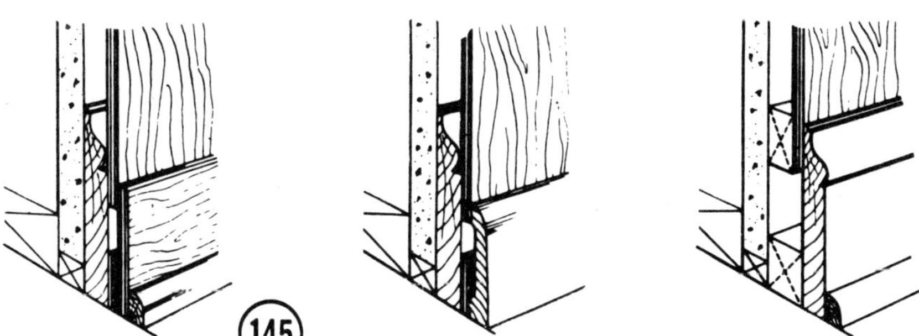

145

Illus. 143 indicates where pros use moldings specified.

Illus. 144 shows type of trim at ceiling while Illus. 145 suggests finishing at base.

OPTIONAL CONSTRUCTION

Illus. 146 shows stairway in the extended house. This plan includes porch area as part of the main house.

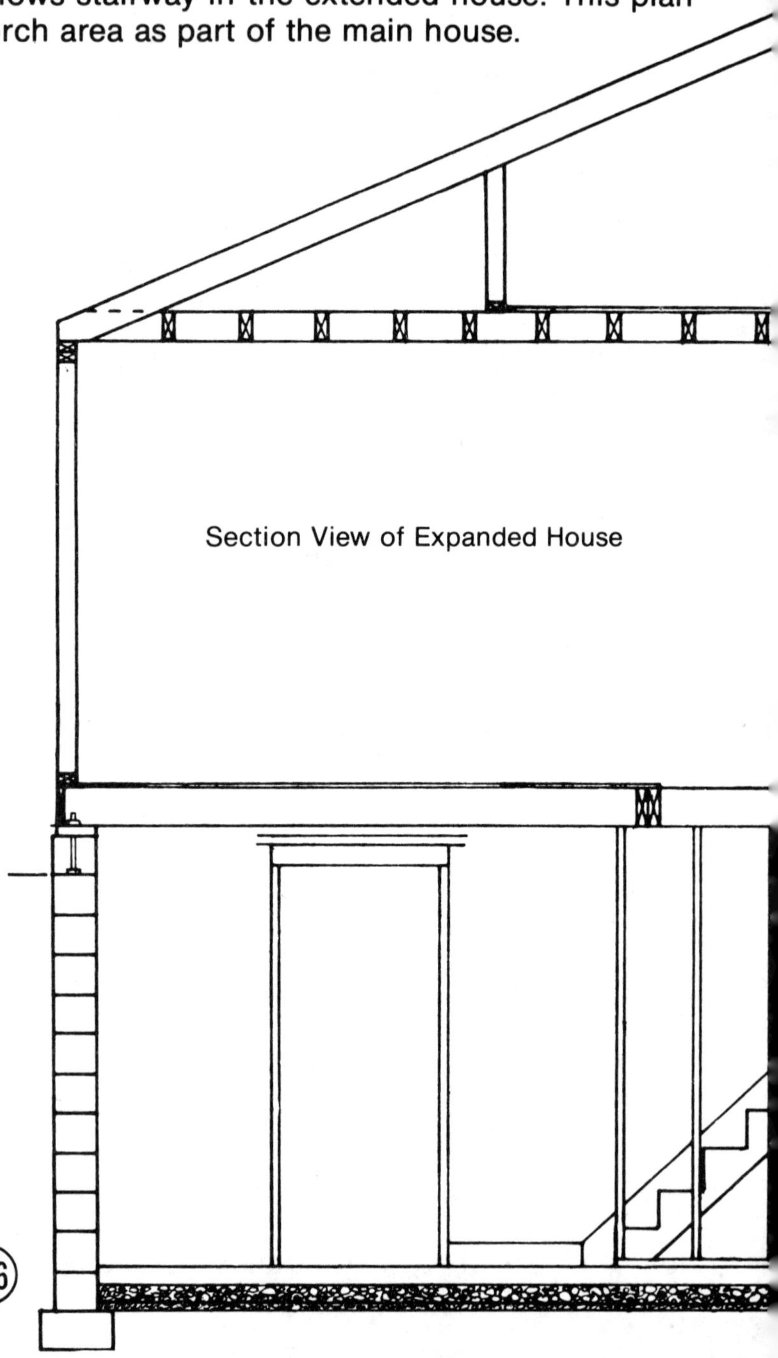

146

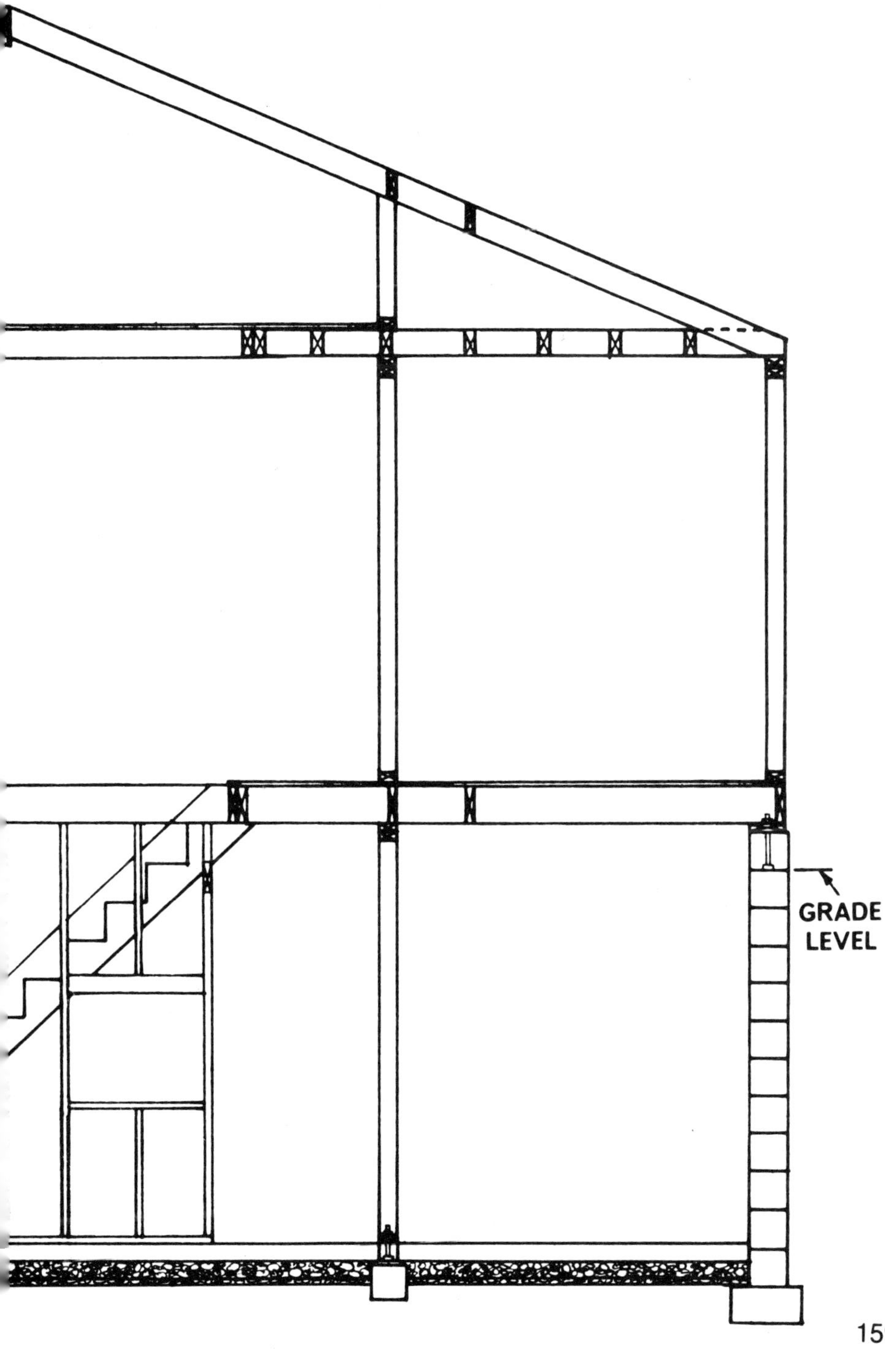
GRADE
LEVEL

SECTION VIEW

EXPANDED HOUSE 2 STORY WITH FULL BASEMENT

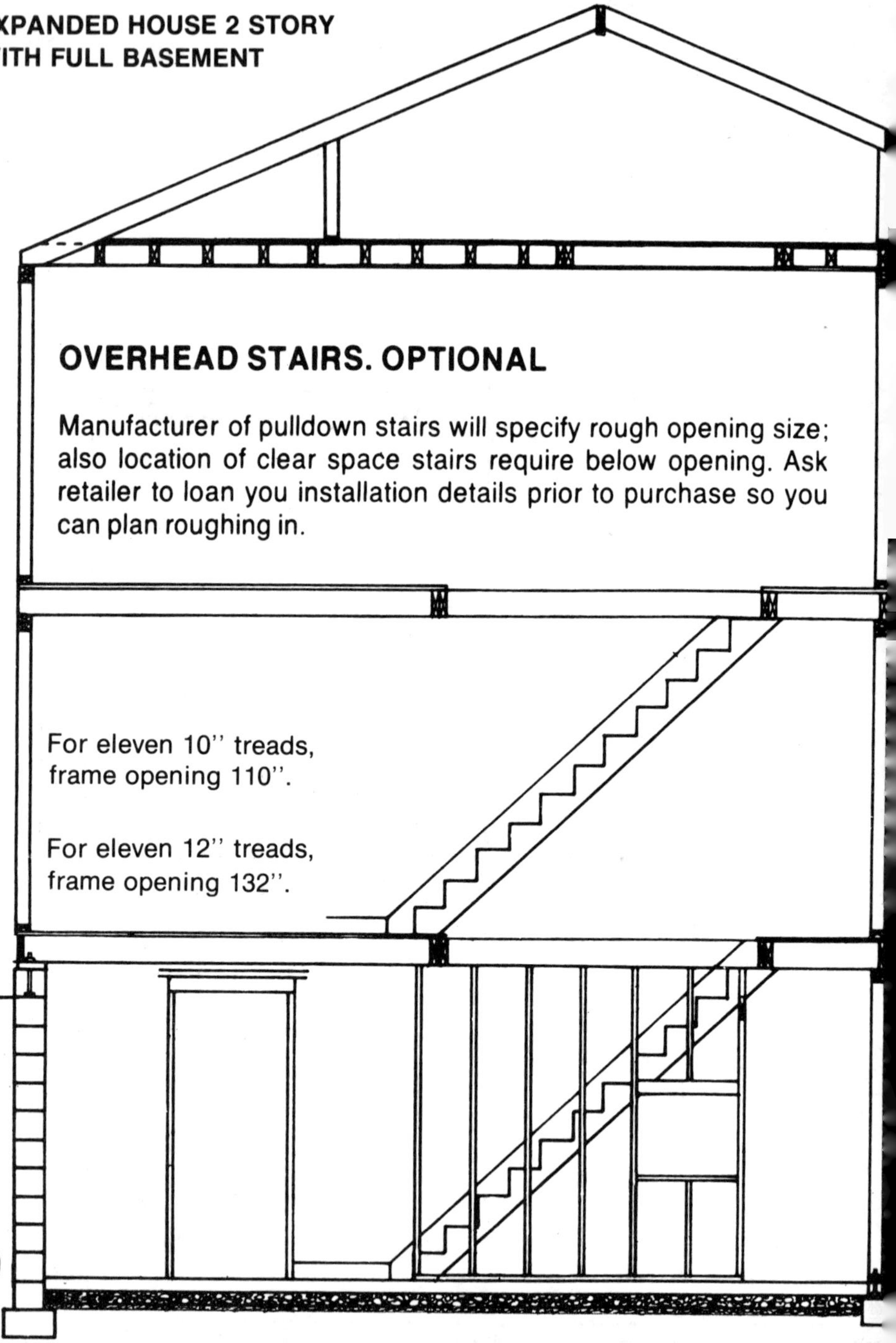

(147)

Illus. 147 shows how a two story could be built over a full size basement, on a slab, or over crawl space.

Illus. 148 A shows how to frame in a second floor with 2 x 8 floor joists if house is extended to include porch area. 2 x 8 is recommended instead of 2 x 6 specified, Illus. 57. Illus. 148B shows how to frame ceiling joists.

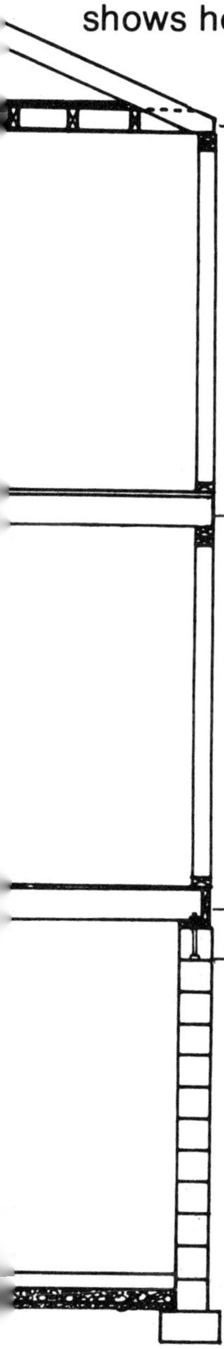

Purchase materials code requires rather than framing lumber specified in list of materials.

FLOOR JOISTS

EXPANDED HOUSE

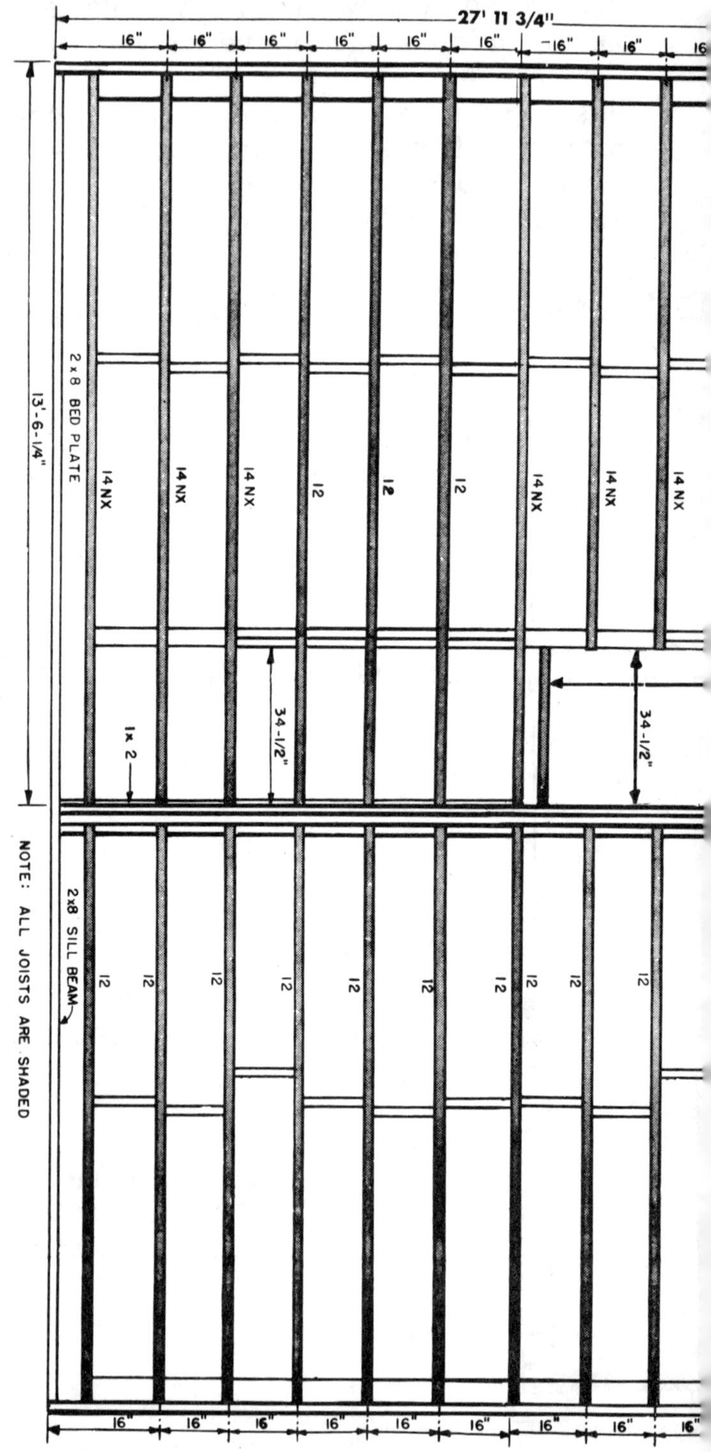

148 A

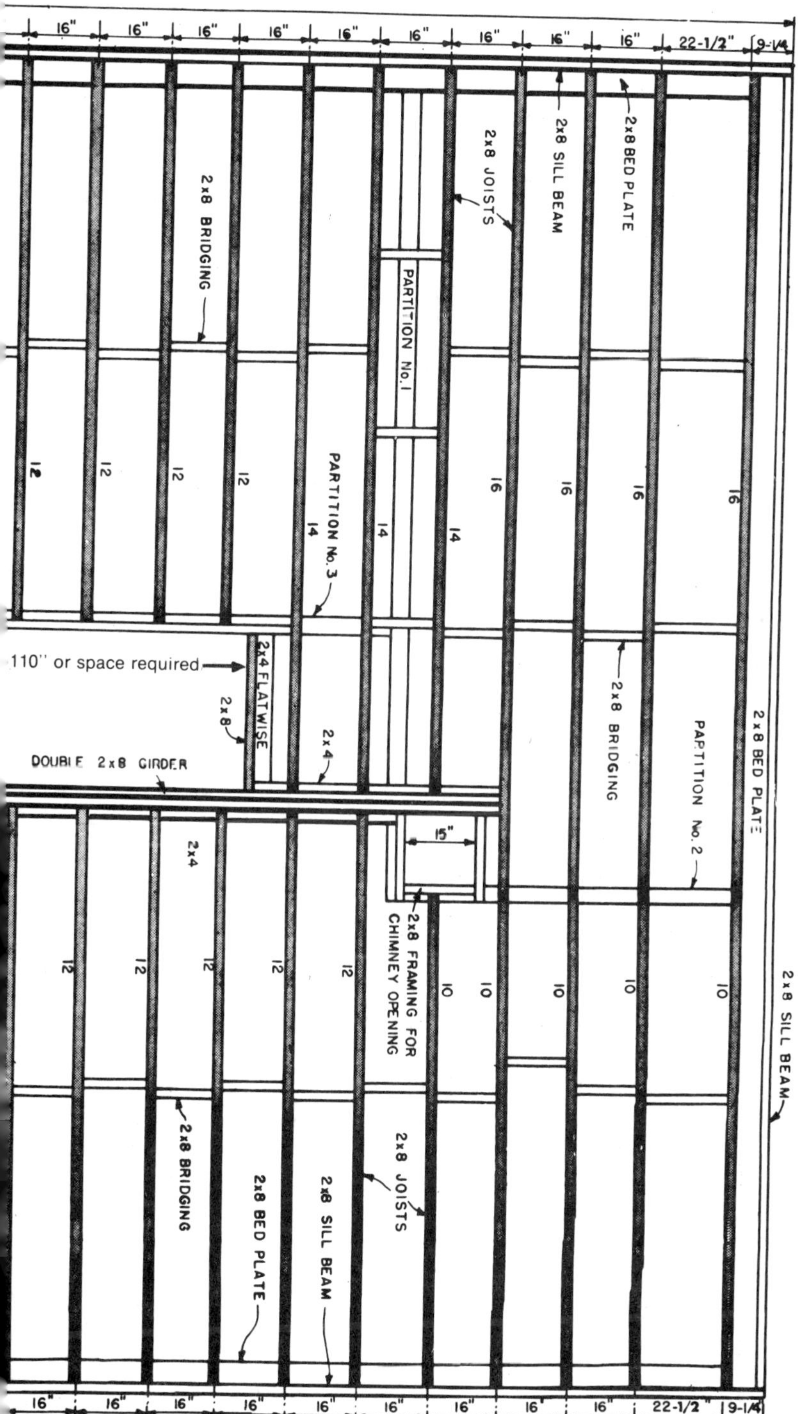
16"
22-1/2"
9-1/4
2x8 BED PLATE
2x8 SILL BEAM
2x8 JOISTS
2x8 BRIDGING
PARTITION No.1
PARTITION No. 3
PARTITION No. 2
12
14
16
10
110" or space required
2x4 FLATWISE
2x8
2x4
DOUBLE 2x8 GIRDER
15"
2x8 FRAMING FOR CHIMNEY OPENING

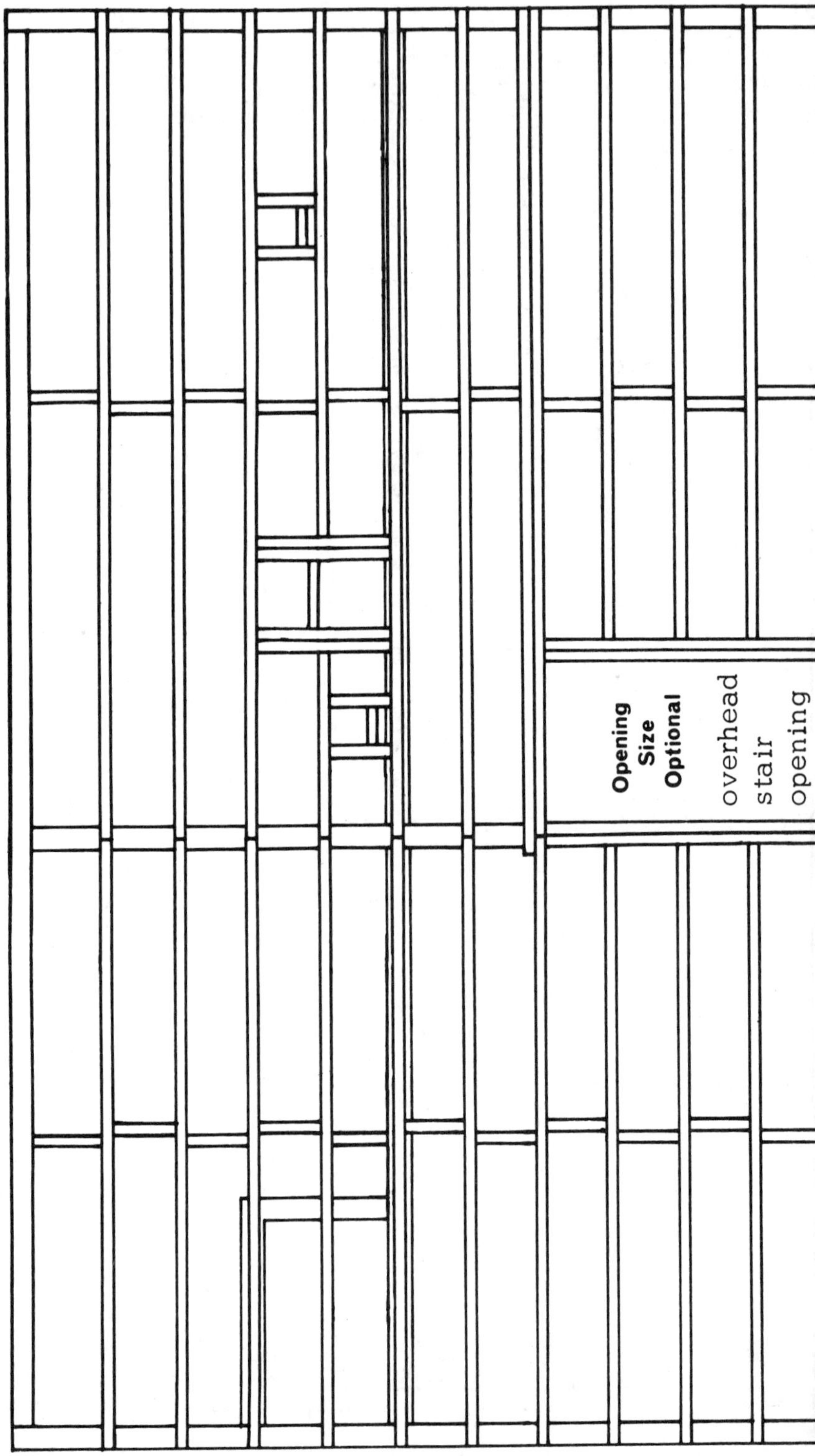

148 B

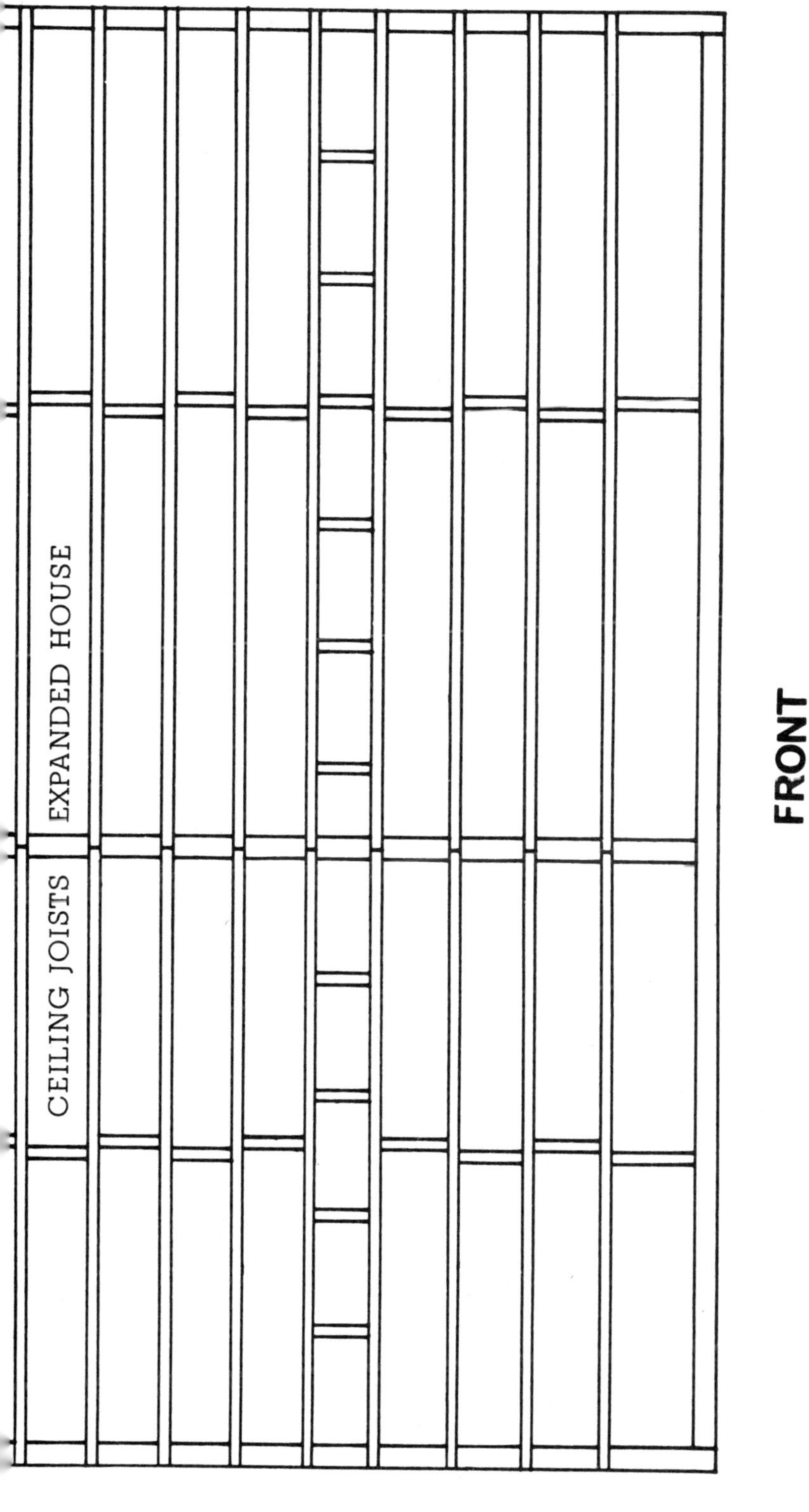
CEILING JOISTS
EXPANDED HOUSE
FRONT

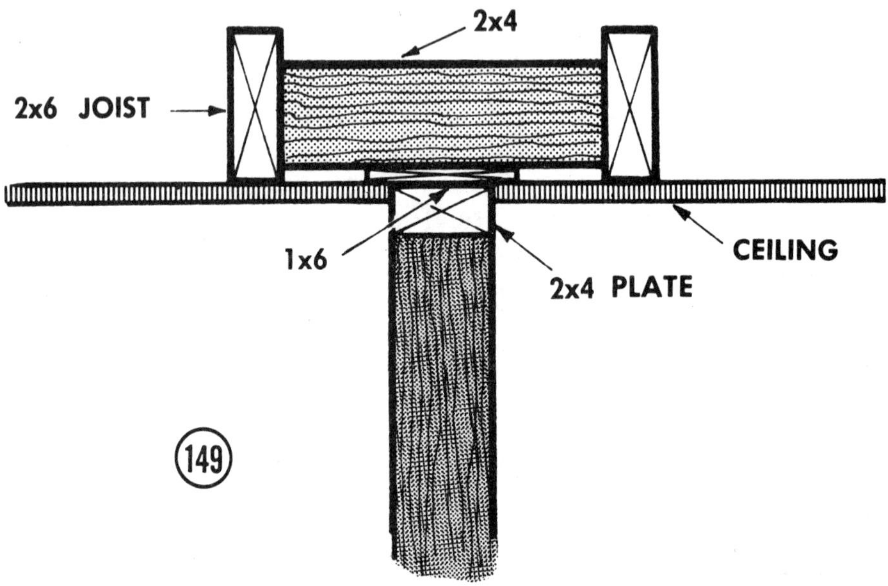

Illus. 149 shows how to nail cats or solid bridging between floor joists when you move a partition. Note 1 x 6. This provides a nailor for ceiling.

Illus. 150 shows cut away of shoe, subflooring, floor joists and sill beam when an outside door is to be installed. When joists run in position shown, nail 2 x 4 cats between joists to support cut edge of subflooring.

Illus. 151 shows method of framing for a second story. Use 2 x 8 floor joists.

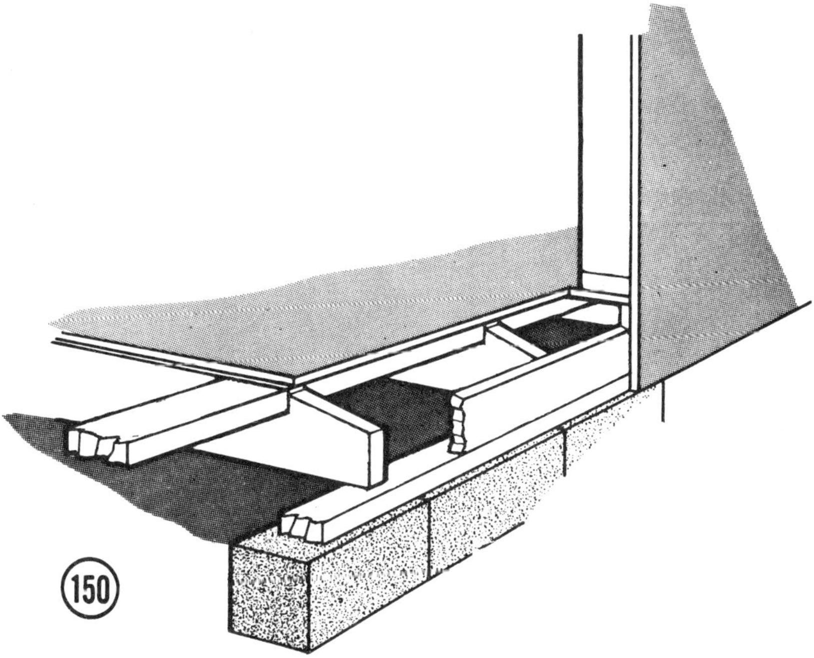

150

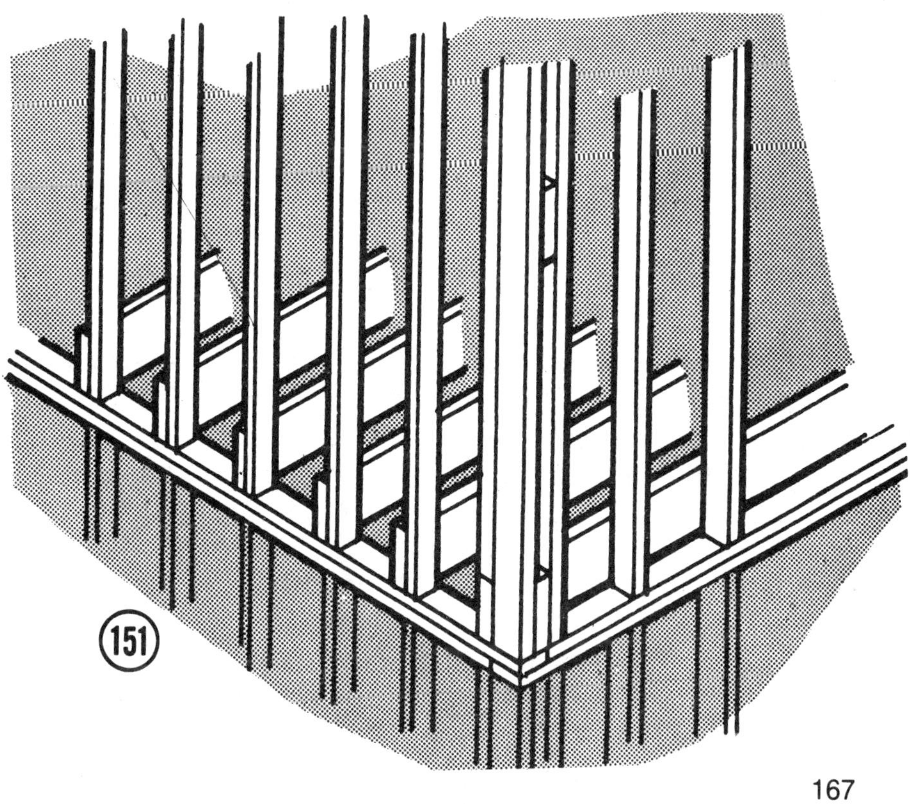

151

TRUSS RAFTERS

You can expedite rafter framing if you buy prefabricated roof trusses, Illus. 152. Ask your lumber dealer about price and delivery. While the truss construction eliminates storage space in attic, it does simplify building.

152

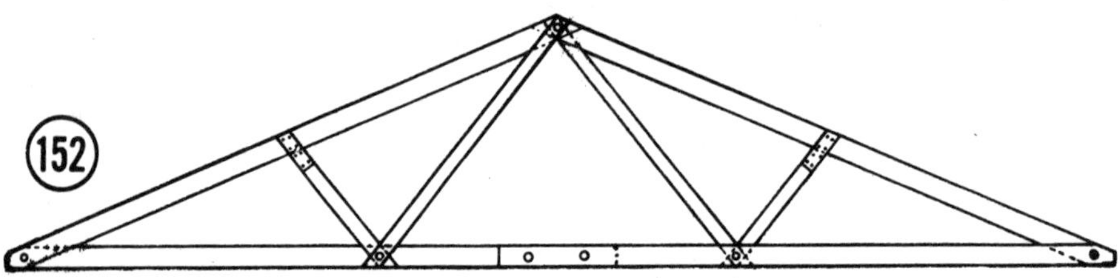

AIR CONDITIONER IN BASEMENT

When the basement rooms are used as a bedroom or recreation room, if local conditions require same, build opening in foundation to rough size conditioner requires, Illus. 153. Install air conditioner in wall as high above grade as possible. If block opening is oversize, use bricks to close opening to size required. This is only recommended when foundation is at least four block height above grade.

Build a 2 x 8 frame to size conditioner requires. Nail bedplate to frame. Install conditioner following manufacturer's directions.

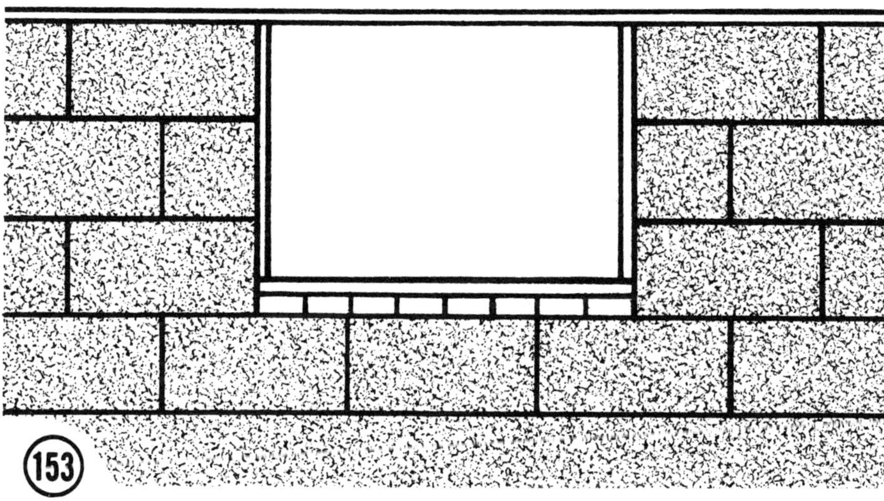

153

STEEL FRAMING CONSTRUCTION

One of the more popular ways of building now utilizes steel columns with either wood or steel beams.

Illus. 8 indicates round footings. These forms are available 12 to 20' in length and in almost every diameter from 6'' up to 4'. Cut form to length required. Plumb and level each. 16 to 24'' diameter will meet almost every code and soil condition.

Dig hole to depth required. Place and level tube. If you run guide lines to indicate top of footing, it not only simplifies placing tubes, but also the bolts needed to anchor plates on columns.

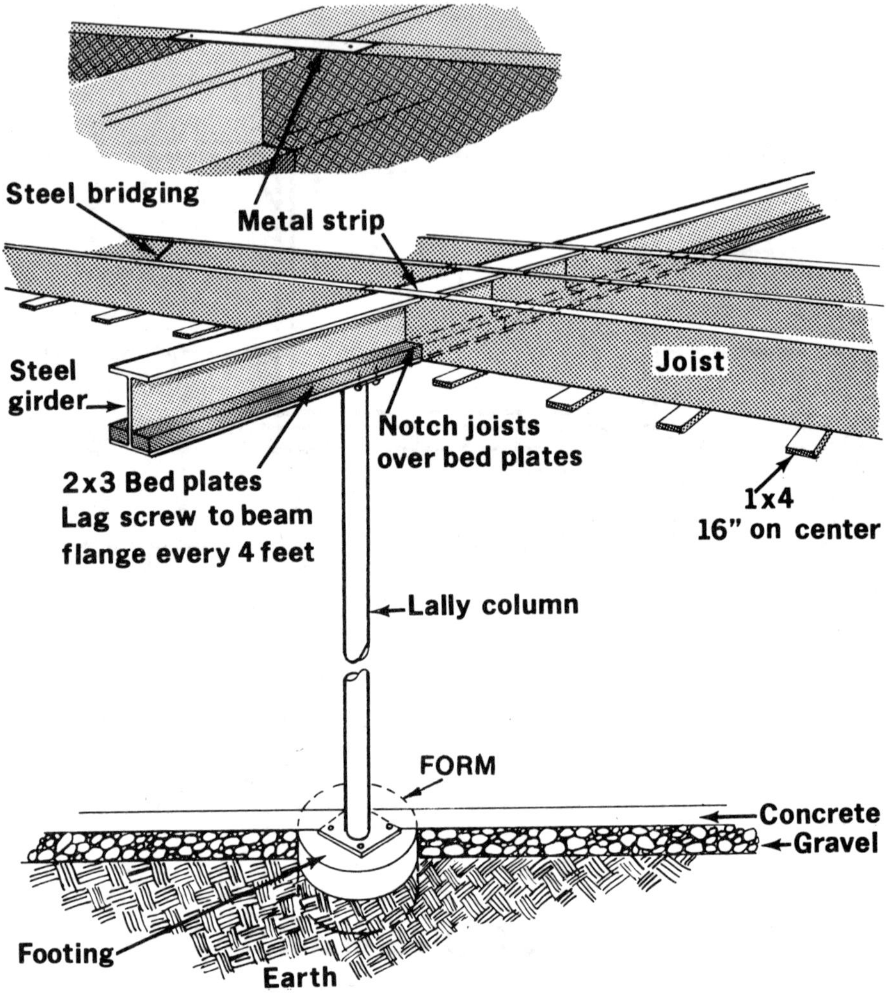
Steel bridging
Metal strip
Steel girder
Joist
Notch joists over bed plates
2x3 Bed plates
Lag screw to beam flange every 4 feet
1x4
16" on center
Lally column
FORM
Concrete
Gravel
Footing
Earth

(154)

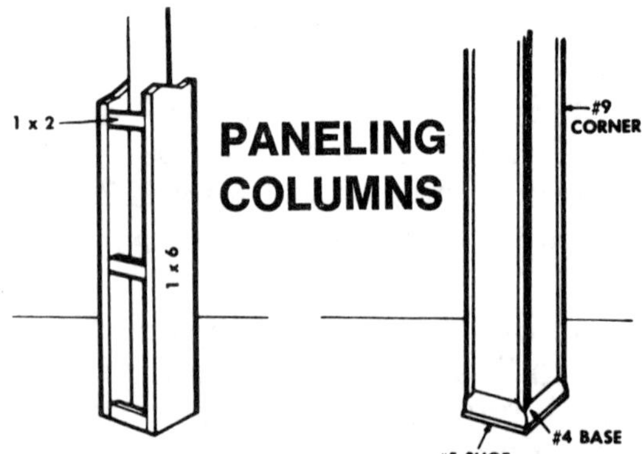
PANELING COLUMNS
1 x 2
1 x 6
#9 CORNER
#4 BASE
#5 SHOE

If you build on a sloping site, buy columns to length required, or lower column in footing. If you embed column in poured concrete, no anchor bolts are required.

Use three 2 x 10 or three 2 x 12 for girder. Buy length required. Always butt joints over a column. If you use 14' lumber, space columns every 7'. If you use 16' lumber, space columns every 8'. If you use a steel I-beam, your supplier can specify column spacing.

Erection of steel columns follows the procedure outlined previously. If steel beams are used, bolt bedplate to steel beam. Notch joists to receive bedplate, Illus. 154. If a wood girder is used, floor joists can be toenailed to girder.

TWO STORY HOUSE

If you plan on building a two story house, Illus. 147,155, use 2 x 8 floor joists instead of 2 x 6 ceiling joists as shown in Illus. 57. Always run floor joists for a second floor in same direction as joists on floor below. This simplifies framing for stairs.

Nail 2 x 8 joists to plate in position shown, Illus. 151. Nail 2 x 4 studs to plate and to floor joists. Ceiling joists, ridge and rafters follow procedure previously outlined.

NOTE: Building a two story house presents no special problems. Partitions should be placed over those on lower floor. If you want to relocate partitions and same runs parallel to joists, but does not rest on joist, nail 2 x 8 bridging between joists, Illus. 61,88.

Always place windows on second floor in line with those on floor below. Since a two story house provides extra bedrooms and considerably more living (and heating) space, you may want to select a different size and type of window than those shown, Illus. 155.

TWO STORY EXPANDED HOUSE WITH PORCH

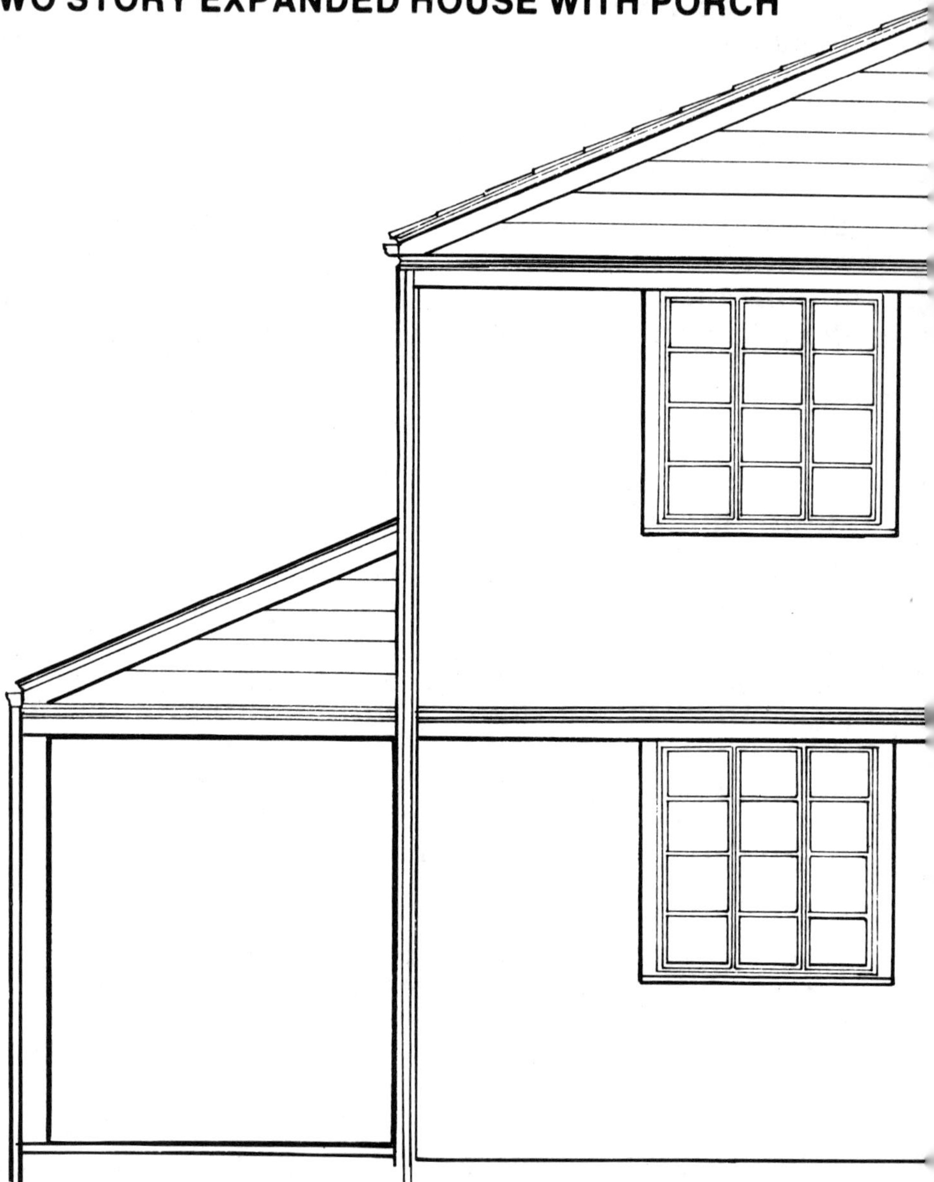

155

FACTS ABOUT WIRING

Today's way of life requires a lot of electricity, so plan adequate wiring. Take time to discuss your needs with the utility company. Explain how you plan on heating the house, what air conditioning unit and other equipment you want to install or may add in the future. Request a 100, 150 or 200 ampere circuit and a three No. 2 wire service entrance.

When estimating Electric Service Entrance Requirements, the following Major Appliances must be considered, individually:

Automatic Washer	700 watts
Built-in Bathroom Heater (how many?) each	1000-1500 watts
Dishwasher-Waste Disposer	1500 watts
Electric Clothes Dryer	4500 watts
Electric Clothes Dryer (High Speed)	9000 watts*
Electric Range	8000-16,000 watts
Electric Water Heater	2000-4000 watts
Electric Water Heater (Quick Recovery Type)	Up to 9000 watts*
Home Freezer	350 watts
Mechanism for fuel-fired Heating Plant	800 watts
Room Air Conditioner (1/3 ton) (how many?) each	750 watts
Room Air Conditioner (3/4 ton) (how many?) each	1200 watts
Water Pump	700 watts
Waste Disposer alone (without Dishwasher)	500 watts

*If the high speed electric clothes dryer or the quick-recovery type water heater is to be installed, it is recommended that more than 100 Amperes' capacity be provided. These appliances, in combination
156 with other equipment, would quickly absorb the available wattage.

Typical wattage of lights and appliances normally used.

LIGHTING	WATTS
Ceiling or Wall (each bulb)	40-150
Floor Lamps (each)	150-300
Fluorescent Lights (each tube)	15-40
Pin-to-Wall Lamps	50-150
Table Lamps (each)	50-150
Ultra Violet Lamp	385
APPLIANCES	
Baker (portable)	800-1000
Bottle Warmer	95
Broiler-Rotisserie	1320 1650
Casserole	1350
Clock	2
Coffee Maker or Percolator	440-1000
Coffee Grinder	150
Corn Popper	1350
Deep Fat Fryer	1350
Egg Cooker	500
Electric Blanket	200
Electric Fan (portable)	100
Electric Roaster	1650
Food Blender	230-250
Hair Dryer	235
Hand Iron (steam or dry)	1000
Heating Pad	60
Heated Tray	500
Ice Cream Freezer	115
Ironer	1650
Knife Sharpener	103
Lawn Mower	250
Mixer	100
Portable Heater	1000
Radio (each)	100
Refrigerator*	150
Sandwich Grill	660-800
Sewing Machine	75
Shaver	12
Skillet	1100
Television	300
Toaster	up to 1150
Vacuum Cleaner	125
Ventilating Fan (built-in)	140
Waffle Iron	up to 1100
Warmer (Rolls, etc.)	100
Waxer-Polisher	350

157 *Each time the refrigerator starts it takes several times this wattage for an instant.

While most utilities are responsible for maintenance from pole to contact on house, if codes permit, use underground cable. Going underground eliminates interruption of service caused by falling branches, ice, etc. The telephone company will also go underground if you provide the trench. If you use city water, the cable and telephone lines can be buried in the same trench.

The amount of amperage needed, like water pressure, requires the proper size service. If you expect to heat with electricity, require complete air conditioning, electric cooking, etc., it's necessary to add up wattage of appliances to see how much house power you require. Check lists, Illus. 156, 157.

It's necessary to feed service from entry into a 100, 150 or 200 ampere main switch and fuse box or circuit breaker, Illus. 158. Install a fuse box that contains the necessary number of circuits. The service panel must match the capacity of the load coming in. It then distributes all or part into those areas where it is to be used. Unless you make provision for an electric dryer, water pump, or even an extra heater in a bathroom, you may find it necessary to run a separate circuit. Another point to consider is the possibility of expanding your house at a later date.

If you build the expanded size house in a cold area, and plan on using electric heat, the utility company will recommend a 150 to 200 ampere service.

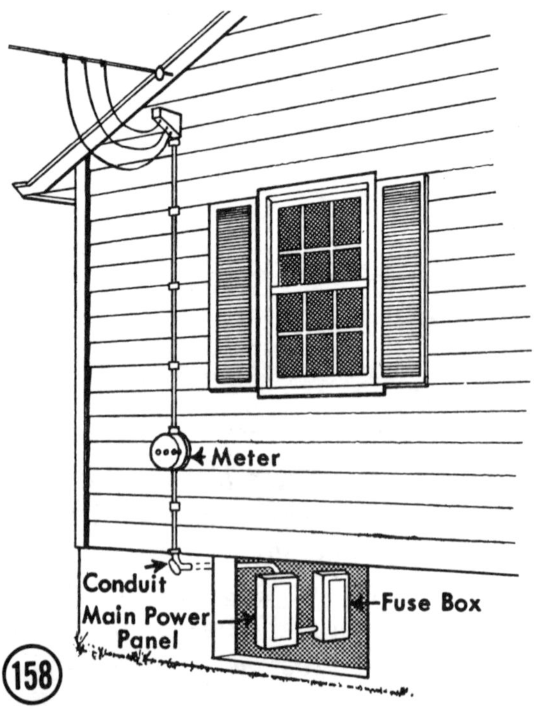

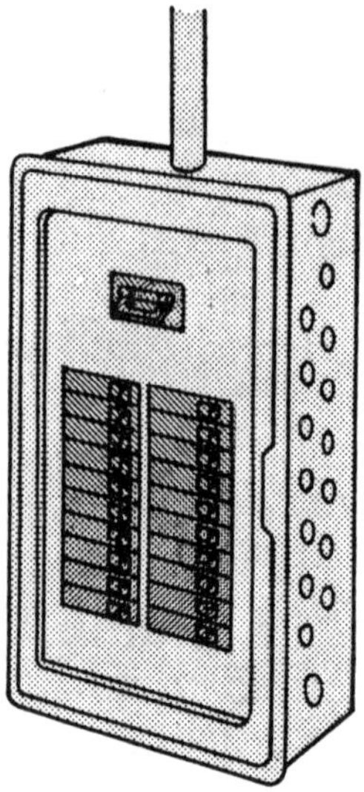

circuit breaker

158

Lights, appliances and equipment can only perform well with needed amperage, so it's necessary to run sufficient circuits.

All wiring is coded by number. The higher the number, the smaller the wire. While #14 wire is acceptable by most codes for general purpose 15 amp fuse circuits, lighting and small appliances, i.e., radio, vacuum cleaner, TV, etc.; separate #12 two wire, 20 amp circuits are recommended for most appliance circuits. A 20 amp #12 circuit should be run to a washer-dryer. Most fuel oil furnaces are serviced by a separate 15 amp circuit.

Air conditioners drawing 1,200 or more watts require a 20 amp #12 three wire circuit. Separate circuits must also be run for a hot water heater, electric range and wall oven.

If you plan on a workshop, it would be well to use #12 three wire to supply a 20 amp circuit for a drill press, table saw, planer, etc.

A 20 amp #12 wire circuit should be run to a deep well pump, and other heavy duty equipment.

To plan intelligently, buy a service panel that contains several spare circuits, so they can be connected when the need arises.

Study Illus. 157 to double check your needs. A good rule to follow provides for one 20 ampere general purpose circuit for each 500 square feet of floor area or a 15 ampere circuit for each 375 square feet.

Never connect all lights or wall outlets in a room on a single circuit. If the fuse blows, you are in the dark. Never connect lights to an appliance circuit.

Illus. 159 shows electric symbols used in residential wiring. Note Illus. 177, 178.

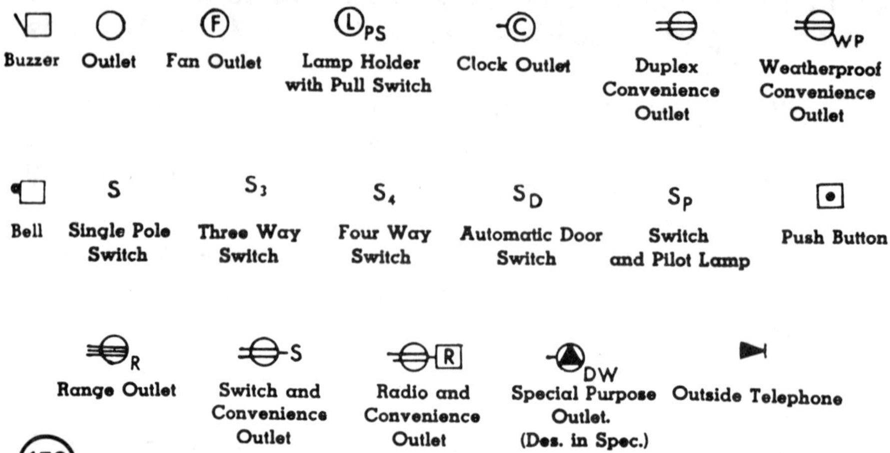

159

FACTS ABOUT A SEPTIC TANK

Before buying property in an undeveloped area, inquire whether the Board of Health will issue a permit for a septic tank. If the property is adjacent to a stream or lake that's part of a watershed or close to wells supplying a community water system, a permit will only be granted if the tank and field can be placed an approved distance. If the parcel doesn't provide the needed space, no permit will be issued. In many areas, a building permit won't be issued unless you obtain a septic tank permit.

Before signing any purchase agreement, ascertain whether a septic tank permit was previously refused. Smart purchasers frequently save themselves considerable heartache by having their lawyer make a land purchase agreement contingent on obtaining both a septic tank and building permit.

Also consider whether the location selected for a tank and field allows sufficient clear space for a house, garage and driveway. Illus. 160 indicates the various size tanks required for 2, 3, 4 and 5 bedroom houses.

160 Recommended Septic Tank Capacities

No. of bedrooms in dwelling	Capacity per bedroom in gallons	Required total tank capacity in gallons	Tank Size			
			Inside width	Inside length	Liquid depth	Total depth
2 or less	375	750	3 ft. 6 in.	7 ft. 6 in.	4 ft. 0 in.	5 ft. 0 in.
3	300	900	3 ft. 6 in.	8 ft. 0 in.	4 ft. 6 in.	5 ft. 6 in.
4	250	1,000	4 ft. 0 in.	8 ft. 0 in.	4 ft. 6 in.	5 ft. 6 in.
5	250	1,250	4 ft. 0 in.	9 ft. 0 in.	4 ft. 6 in.	5 ft. 6 in.

Prospective property buyers who plan on building outside of a sewer district should study the installation of a septic tank and field, as it's important in making the property usable. Most rural communities adjacent to any size city only permit installation if it's placed 5' or more from the house, at least 50' from a cistern, 100' from a well, and that the field is at least 10' from property line, Illus. 161.

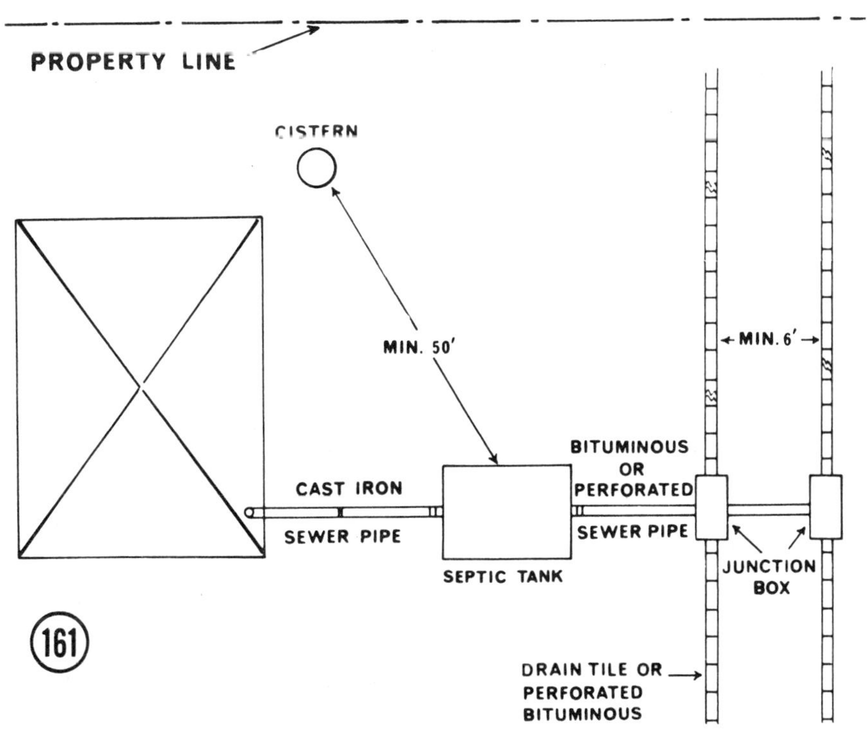

161

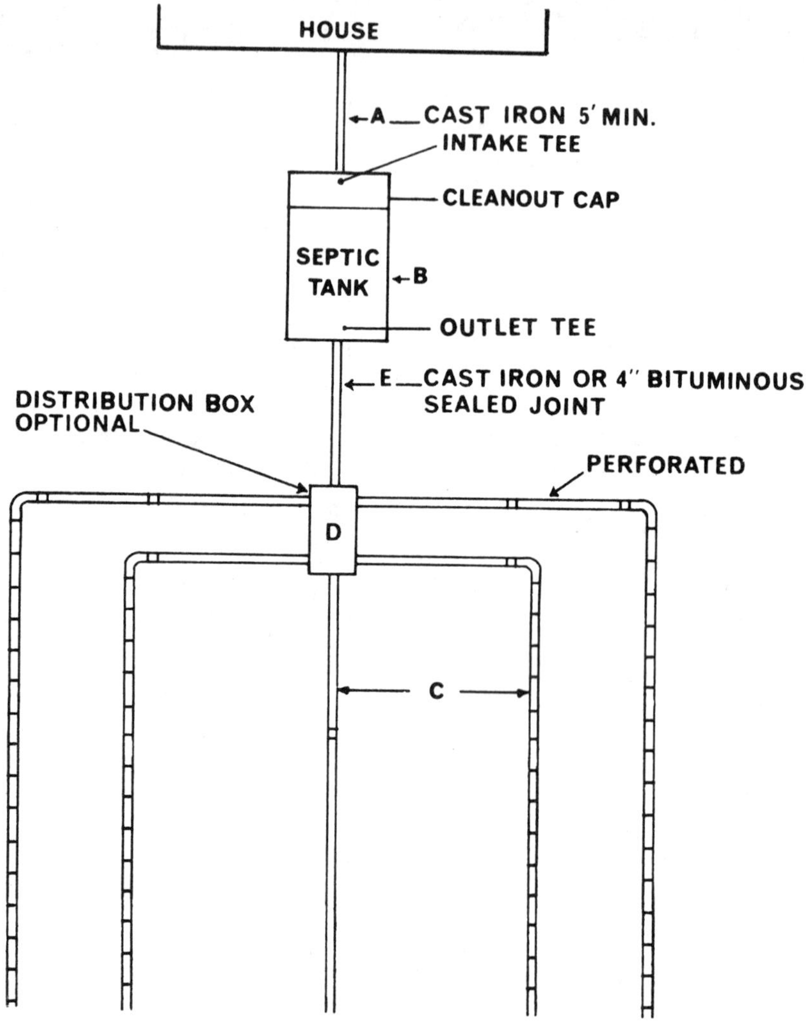

162

When selecting a site, be sure to consider any well, cistern or stream on your neighbor's property.

As noted in Illus. 161,162, codes specify hub and spigot or hubless cast iron sewer pipe between house and septic tank. This must be 5'0" or longer. Cast iron or bituminous sewer pipe is acceptable between the septic tank and either a junction or distribution box.

The sewer line between house and septic tank must slope at least ¼ to 1" per foot, as does the connection between the tank and the junction box. Follow local code requirements. The septic tank must be level.

The area required for the field is determined by the number of lineal feet the field requires. You can run this in two, three or as many rows as shape of space permits. Most codes specify a 6' minimum spacing between fields.

A septic tank and field should never be located in a low, wet area or on rock, and only in an area where no tree roots, shrubbery, etc., will send roots down to the tile. After selecting a site, call the Department of Health and they will make a free inspection or will advise what must be done to pass inspection.

Most inspectors not only inspect the site, make inquiry concerning the number of bedrooms the house will contain, recommend size of tank, shape and length of field, but will also make an absorption test to ascertain how fast the subsoil will absorb water. Or they will advise you how to make a test.

While installing a system is relatively simple, a septic tank installation can become real sticky when you learn that such and such a septic tank company can obtain a permit for you, then quotes an exorbitant cost.

Regardless of what an "in" septic tank installer may tell you, no one can stop you from making your own installation, providing you follow established regulations.

If you think anyone in the health department is giving you a hard time, there are two good reasons why this may be so. Most health departments make an honest and determined effort to protect the area adjacent to watershed property. Regardless of whether it's a nearby stream, runoff into a stream, or a well on your property or neighboring property, the field must be laid according to regulation. The second reason is financial. If a local septic tank company has a monopoly, they may overcharge.

If you read and learn how to install a septic tank before you start talking business to an installer, you will save yourself a lot of time, money and aggravation. If you then decide to do part or all the work, you can save even more money. Either way, you get a better job when you know what needs to be done.

A septic tank system consists of a large metal or concrete holding tank B, Illus. 162; a distribution center, D; and a field of 4'' diameter perforated drain tile, C. Many codes permit installation without a distribution box. They allow junction boxes as shown in Illus. 161.

The sewer line E, Illus. 162, must be 4'' hub and spigot or hubless cast iron, 5'0'' or more in length. Hubless 4'' cast iron and/or plastic pipe is also approved by most health departments.

You can buy a metal or prefabricated concrete septic tank, or build a 6'' form and pour the concrete tank in place. Reinforce sides and bottom with ½'' rods, or use 6 x 6'' reinforcing wire.

Since tanks purchased readymade will be delivered by a truck with a crane that positions tank level in the excavation you have prepared, the question you must resolve is whether the site selected for the tank can be reached by a truck. Give consideration to this simple but important fact since once you start excavating for a house you might not be able to truck and crane a tank into position.

If you prefer to pour a tank in place, will a ready mix truck be able to come close enough to chute or wheelbarrow the ready mix? Or does the location of the site automatically require you to rent a concrete mixing machine and do the work on the site? With planks properly placed to form a level or downhill walk, two or three willing souls can usually wheelbarrow concrete from a ready mix truck fast enough to pour a tank without paying overtime for a waiting truck. If you decide to buy ready mix, ask the company how long you can hold the truck on the job before overtime sets in.

Always check tank with a level after it's placed in position and before the truck leaves.

Trenches for a field are usually 18" wide and to a depth that permits 4 to 6" of gravel on bottom, 4" tile, plus 2 to 4" of gravel on top, Illus. 163. The 4" drain tile used in the field must be placed at a pitch codes specify. This could be 2 to 4" in 100'.

(163)

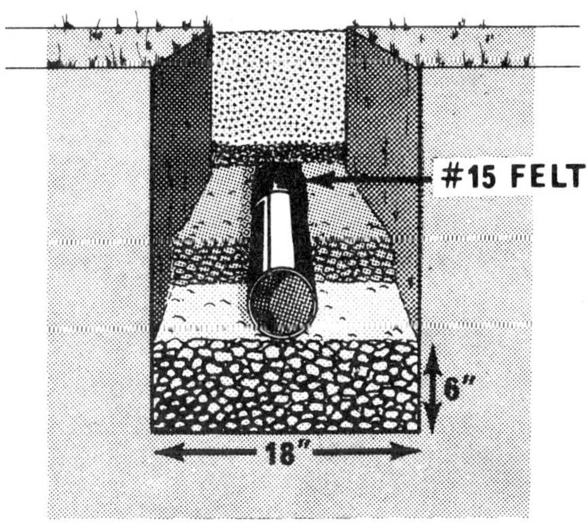

Building a form and pouring a septic tank, or laying block for a tank to size local codes require, is no big deal. Book #697 Forms, Footings, Foundations, Framing, Stair Building provides complete directions that cover renting, as well as building forms. If you decide to buy a concrete or metal tank (providing codes approve same), buy a size that will accommodate your present needs as well as any future plans you might have for adding an extra lavatory or bathroom. There's no economy in putting in a minimum size when a larger one only costs a little more.

Those who contract a complete septic tank and field installation should obtain a written agreement specifying when the work will be done and that payment will only be made after the installation has been approved by the health department.

If you decide to do the work yourself but wish to buy a concrete or metal tank, make certain the truck delivering same can lower the tank into the excavation you have prepared. This requires digging to depth tank requires and leveling bottom. Use a straight 2 x 4 and a level to check bottom of excavation. The depth and size of hole for a tank is determined by its size and shape. Only lay A, E, D and C, Illus. 162, after tank is in position.

Since the waste line A from house to the tank must pitch ¼" per foot, or pitch local codes require, the depth of the hole will be determined by the intake TEE, Illus. 164. Dig trench from house to selected site. Slope trench to pitch pipe requires. When 5'0" or further away, place a TEE, Illus. 165, in position on the end of the pipe. To estimate depth of excavation, measure from bottom of TEE to depth below TEE your tank requires. Note position of inlet TEE, Illus. 164.

Codes recommend outlet TEE be placed 3" lower than inlet.

When installing a new tank, always install a cleanout plug, Illus. 166, in direct line with the inlet TEE. This greatly simplifies servicing a septic tank.

When installing a septic tank on rolling or hilly land, the field must only slope amount per foot local codes specify. This may require filling in low areas. A septic tank for a "foundation house," Illus. 6, must be lower than floor level. The field must slope away from house.

Many codes permit building a septic tank with concrete block. Excavate to size and depth tank requires. Level bottom, then dig a 10" wide footing trench, 4" lower than floor, around perimeter of floor area. Pour footings and allow them to set. Lay a thick bed of mortar and position a starter course of 10 x 8 x 16 or 18" blocks. These should project 4" above floor area.

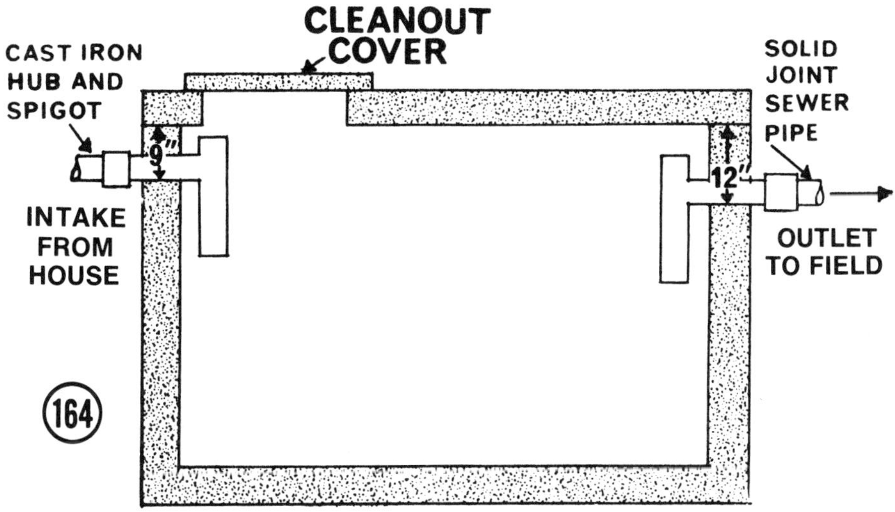

164

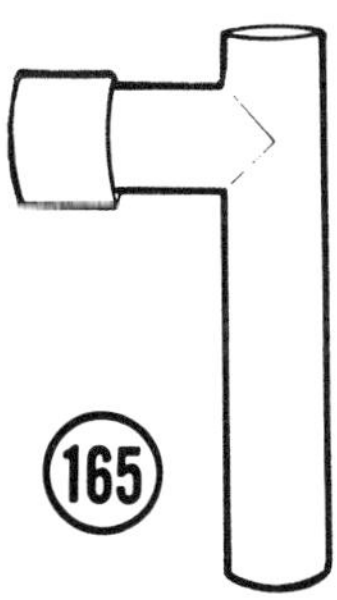

165

166

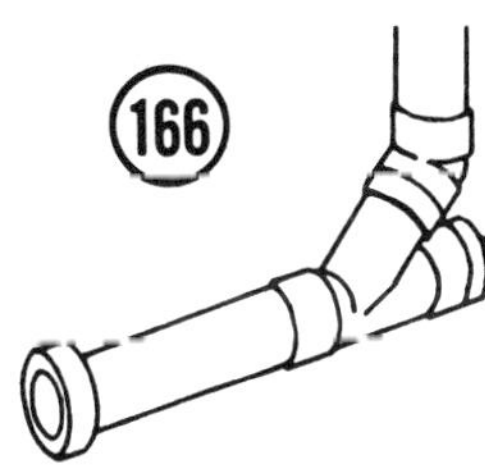

CAST IRON PIPE TO SEPTIC TANK

Cut 6 x 6" reinforcing wire to size floor area requires. Raise wire about 1" and pour floor flush with top of block. Fill core of blocks with concrete. Allow slab to set three days, then build walls using 8 x 8 x 16" block. These are placed flush with inside edge of 10" block. Fill core of the first three courses of block with concrete. If you are building a large tank, lay reinforcing rod, Illus. 22, in every third course.

Set Inlet TEE and Outlet TEE in position shown, Illus. 164.

Allow tank to set at least five days, then paint inside and outside with hot tar or asphalt cement.

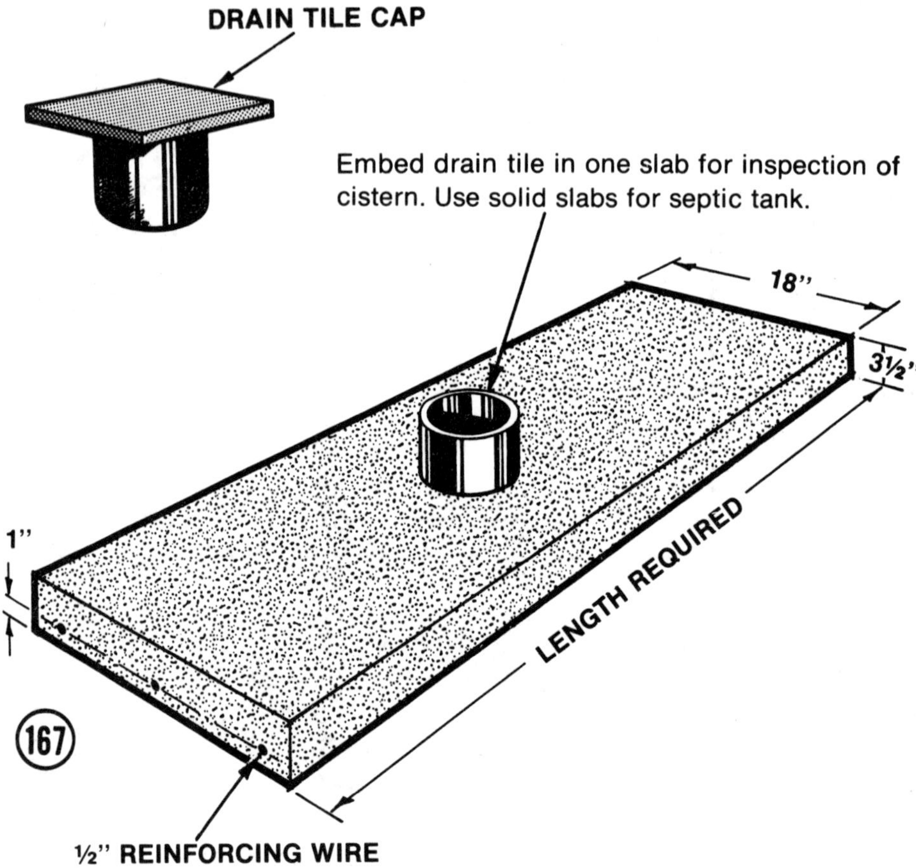

(167)

Following local code, lay field to depth specified. Cover with size and depth of gravel specified, then cover gravel with #15 felt before backfilling. Since the Board of Health will want to make an inspection before you cover the field, follow directions they provide and under no circumstances give the inspector a hard time.

Cover tank with precast slabs, Illus. 167.

When laying perforated bituminous field tile, follow code recommendations regarding a distribution box, Illus. 162, or junction boxes, Illus. 161.

THE FOUNDATION HOUSE

Within a year after the first edition of this low cost house was first released, we began to receive letters from newlyweds, retirees and singles who enthusiastically reported how they had built and enjoyed living in a foundation house. Some described how they had expanded the basement to include the exact size space shown on the first floor. These early "earth sheltered" houses had much to offer. The living space was less costly to build, heat in winter and cool in summer.

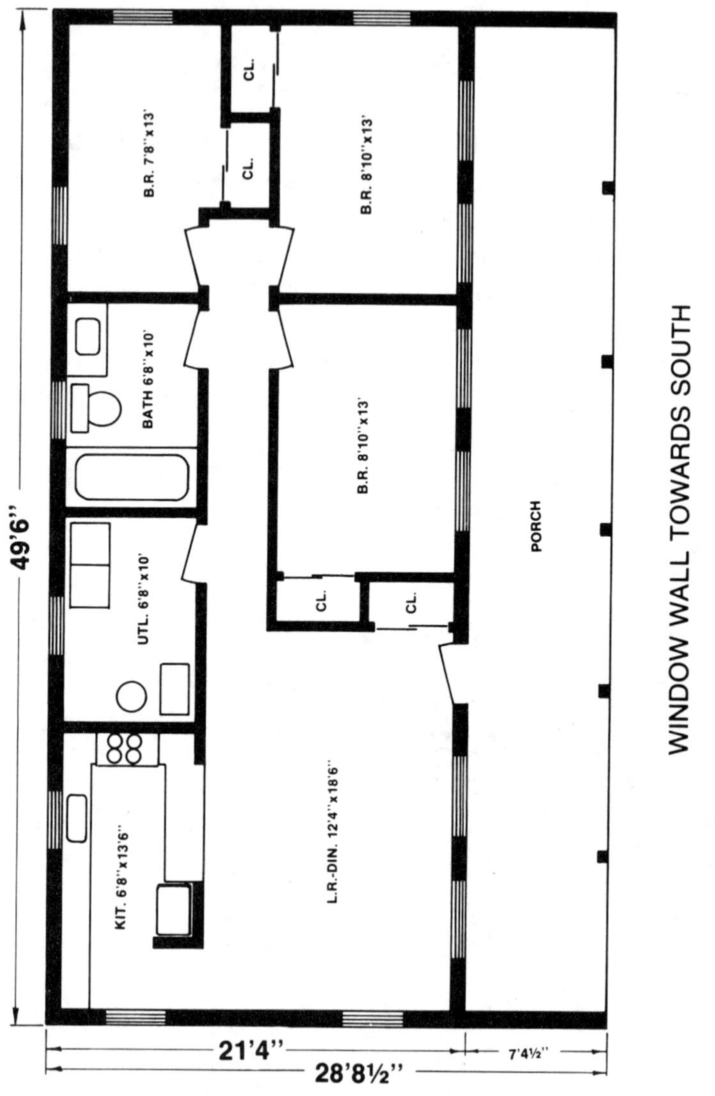

FLOOR PLAN FOR FOUNDATION HOUSE

While the first printing referred to the economy of living "below grade," the book did not provide detailed directions. Many letters mentioned reading Book #697 Forms, Footings, Foundations, Framing, Stair Building and #615 How to Modernize a Basement to find answers to various construction details. Backfilling a side hill house with earth on three sides saves a lot of energy. Most foundation houses can be heated with a woodburning stove alone.

Building a side hill foundation house, like all earth sheltered dwellings, requires a waste disposal tank and pump that can carry waste to a septic tank and field. The field must be at a distance and level that permits waste to penetrate soil without filtering back toward foundation. Since a disposal tank and pump are frequently needed by houses with bathrooms in a basement, your local Board of Health will recommend size of tank and pump required.

Illus. 168 shows a cutaway section. Note gravel alongside footing. 4" tile, placed end-to-end along entire footing, pitches toward a dry well. The dry well must be located at a lower depth and sufficient distance from foundation to carry off all seepage. Wrap tile joints with strips of #15 felt. Water seeps into joints and is carried to dry well.

To relieve external stress, pour steel reinforced concrete buttressed walls for foundation house constructed at the base of a steep grade, Illus. 168. Build to overall height required.

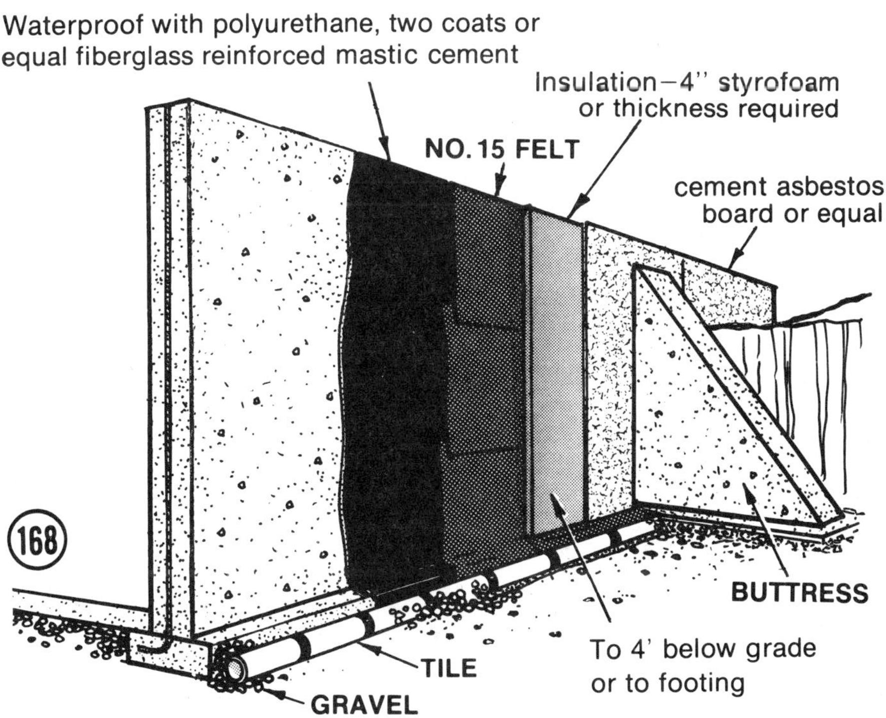

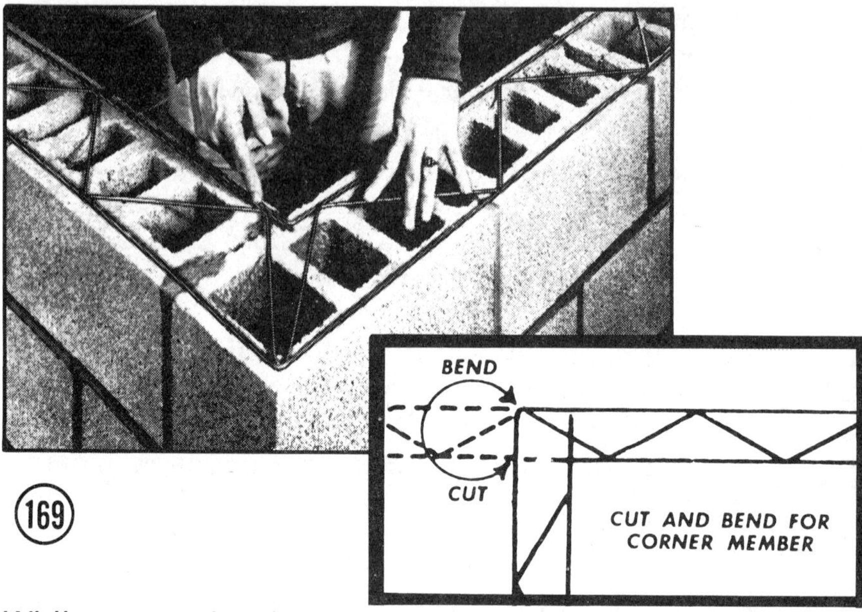

169

While a steel reinforced poured concrete footing and foundation is recommended, if conditions permit, blocks with reinforced mortar joints, Illus. 169, can be used. Embed wire in mortar bed. Your masonry retailer will recommend adding waterproofing to mortar used to lay blocks and seal exterior.

To waterproof a concrete block wall, add required amount of waterproofing compound to water prior to mixing one part cement to two parts sand. Trowel a ⅜ to ½" plaster coat to block, Illus. 170. Before it sets, scratch it, Illus. 171. Allow to dry, then apply a second coat, Illus. 172. Spray the surface once a day with a fine mist for three to four days. This helps cure concrete.

170

171

(172)

(173)

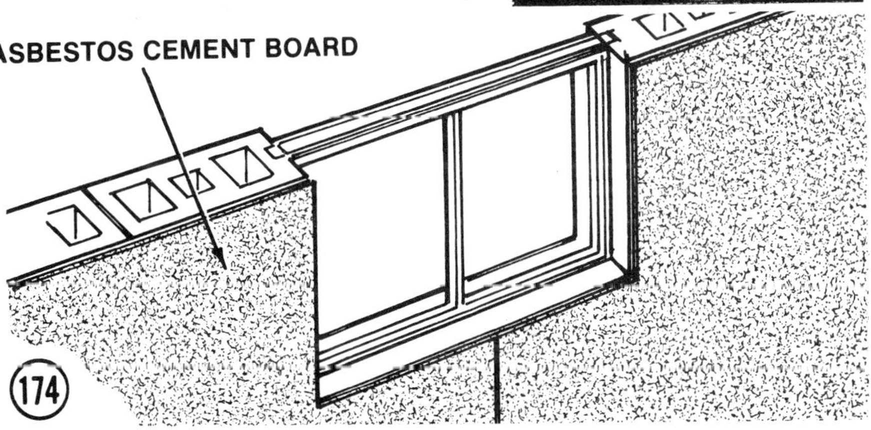

(174)

Set anchor bolts; fill top core of block with concrete, Illus. 39. Install ceiling joists, rafters and insulation following directions outlined previously

Allow plaster coat to air cure for two weeks before applying asphalt roofing or hot tar, Illus. 173. Brush on one coat. Allow to dry, then apply second coat. While wet, embed a layer of #15 felt horizontally. Overlap each course 4 to 6" in a bed of asphalt. Start at bottom and work up. Allow starting course to completely cover footings prior to laying gravel and 4" tile. Cover felt with asphalt or hot tar and another layer of #15 felt prior to application of rigid insulation board. Use thickness local climatic conditions require. The entire wall is then covered with 4 x 8 sheets of asbestos cement board, Illus. 174, to protect insulation.

Use care when backfilling against cement asbestos board and insulation. Keep all rocks at least a foot to 18" away from side of foundation.

Book #697 Forms, Footings, Foundations, Framing, Stair Building provides much helpful information on the installation of wood windows both in concrete block and in a poured concrete foundation. It also explains how to lay a felt or polyethylene vapor barrier over a level sand or earth floor prior to laying concrete; how to install sleepers in concrete; the need for expansion joints, sealing a below grade floor against moisture.

For those who want to expedite construction, it explains the availability of prefabricated forms, installation of steel rod prior to pouring ready mix.

Install anchor bolts in top course, Illus. 40, a 2 x 8 bedplate, ceiling joists, rafters and roofing following procedure previously described. Apply insulation between ceiling joists (use thickness local conditions require), apply #15 felt to gable ends, siding, etc.

Those building a foundation house into a side hill where a side may come in contact with the earth should use precast concrete joists, Illus. 175. As your masonry supply retailer will confirm, these are placed on the foundation block. No wood plate is used. If you want to install a window, it must be installed under a reinforced lintel as explained in Book #697. Always use length of joist that spans area between outside wall and a 4 x 8 x 16" concrete partition wall.

Steel reinforced concrete joists are available in many lengths. Discuss your needs with a local supplier. He will deliver same and place each in position with a crane (providing the truck has access to the foundation).

Those building an earth sheltered house use the precast joists over entire foundation. As previously stated, this necessitates building all partitions with 4 x 8 x 16" concrete block to provide needed support.

(175)

Precast concrete slabs, usually 2" thick, 2' square, are laid between joists. A precast shoulder on joist receives these slabs. Pros usually fill space between joists with 4" block. A polyethylene sheet or #15 felt is embedded in asphalt. A concrete slab is poured in two stages. After pouring about 2", 6 x 6" reinforcing wire is laid in wet concrete. All concrete used is usually treated with a suitable quantity of waterproofing compound. Or 6 x 6" reinforcing wire can be spread over entire area before the first pouring. The reinforcing wire is then raised above slab about 1½ to 2" before pouring balance of concrete.

When the concrete roof has been allowed to cure the time prescribed by manufacturer of waterproofing, mop deck with asphalt roofing, then lay a course of roofing felt. Overlap courses of felt at least 12 to 18", then mop on another layer of asphalt and cover with another layer of #15 felt. This is then covered with asphalt and rigid insulation board is embedded in asphalt. Carefully cover roof with earth, no stone, to thickness climatic conditions require. Ask retailer for instructions covering any material he recommends.

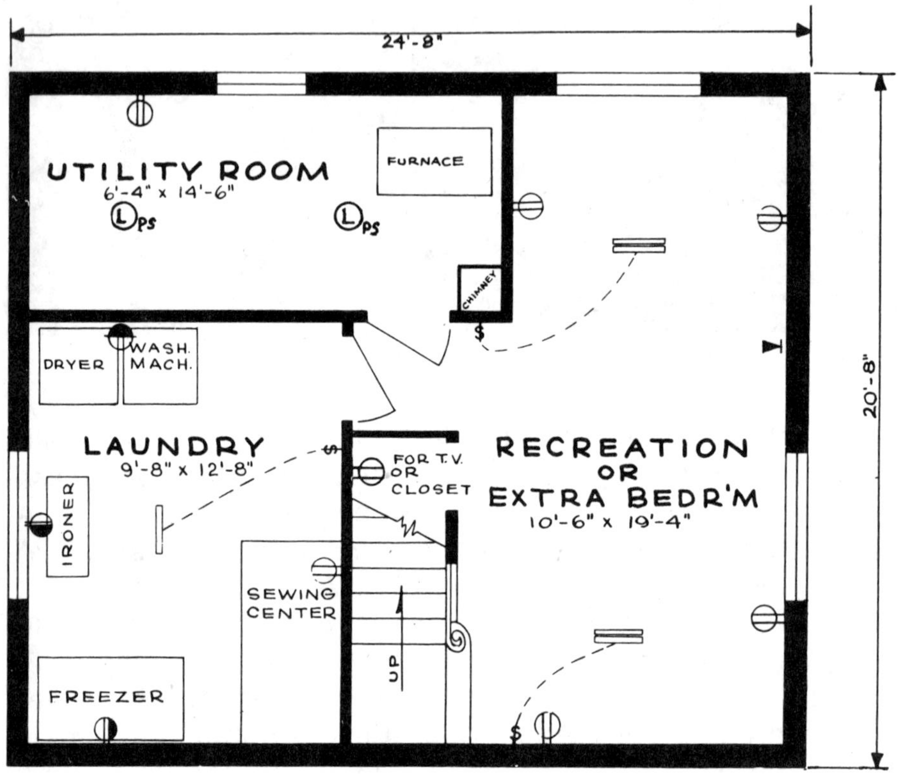

176 **WIRING PLAN BASEMENT**

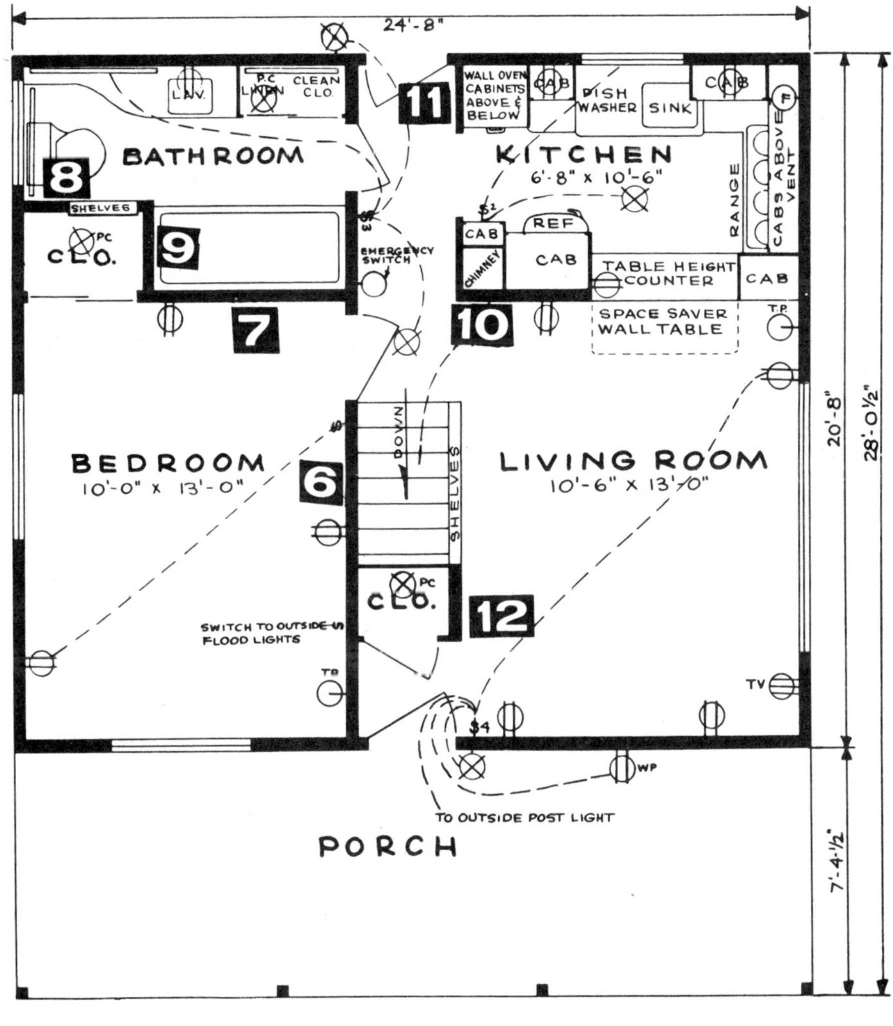

(177) WIRING PLAN FIRST FLOOR

OPPORTUNITY UNLIMITED

The nation's economic system is undergoing a traumatic experience. As unemployment increases, crime rises. All turmoil creates FEAR and fear creates opportunity for those willing to listen, act and then expend needed effort.

Building a house, regardless of size, creates shelter at a cost you can afford. It also creates a new you and unlimited business of your own opportunities. Overcome fear, build one house and you are no longer fearful of building a second, third or a hundred more after you obtain needed experience.

The first step is the hardest one to take. Lack of confidence or willingness to try can keep you off first base. Take a shovel and dig a foundation and you begin to expand your sphere of activity, plus start a way of life you previously only dreamed of living.

Decades before bulldozers were created, men with picks and shovels dug out basements for thousands of homes. These men, many with no formal education, became industrial leaders.

As the automotive and other heavy industries turn to computerized robots in the fabrication of parts formerly assembled by human hands, the jobless rate will continue to increase. Your future depends on what you are willing to do to solve your own employment problem. The step-by-step directions in this book and the others in this series describe and graphically illustrate every step. And each helps rebuild your self confidence. If you can read and follow direction specified, you can do everything the author suggests.

All Easi-Bild Books are easy to read. Each can help a school dropout as easily as they can help relieve tension for the executive out of work.

The key to your future rests solely with WHAT YOU DO. TRY and you will succeed.

CONCRETE BLOCK SIZES

A– 7 5/8" B– 15 5/8" C–9 5/8" D– 11 5/8" E– 3 5/8"

APPROXIMATE METRIC SIZE A-19.37 B-39.7 C-24.45 D-29.53 E-9.21

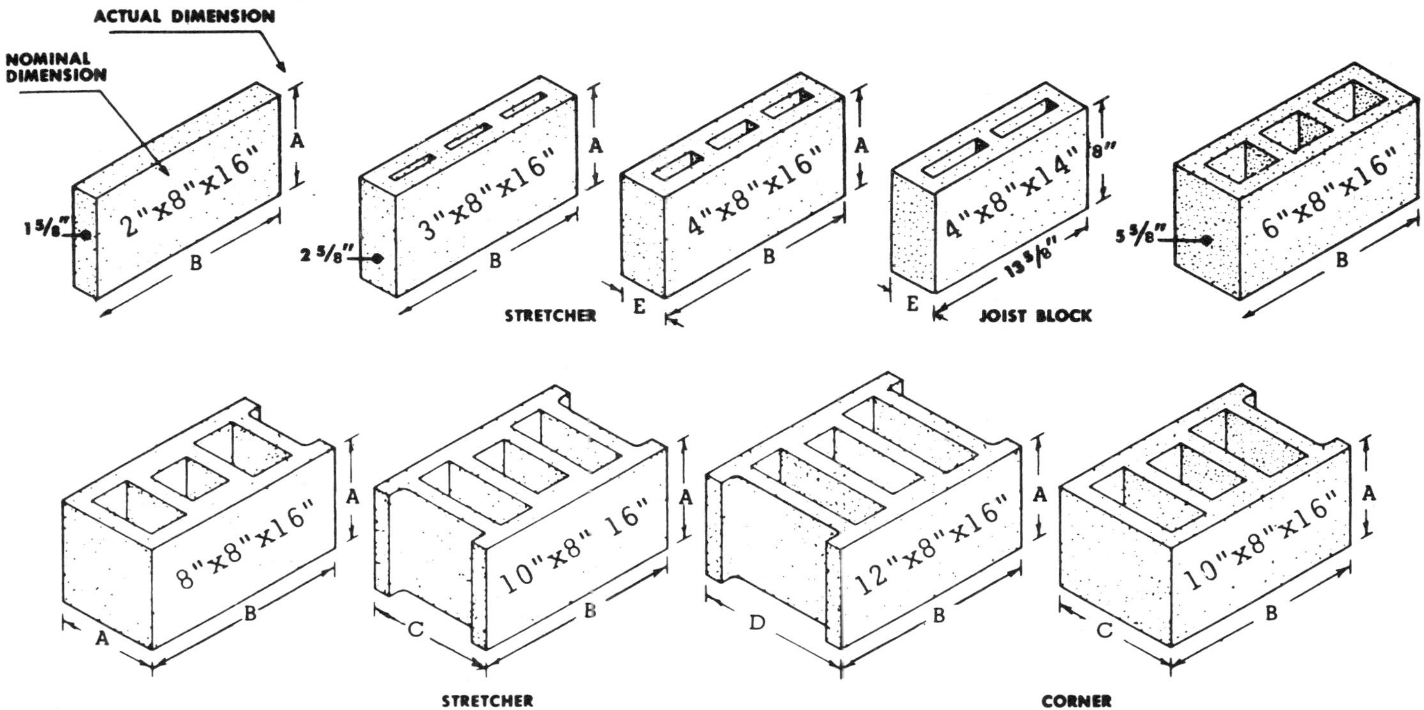

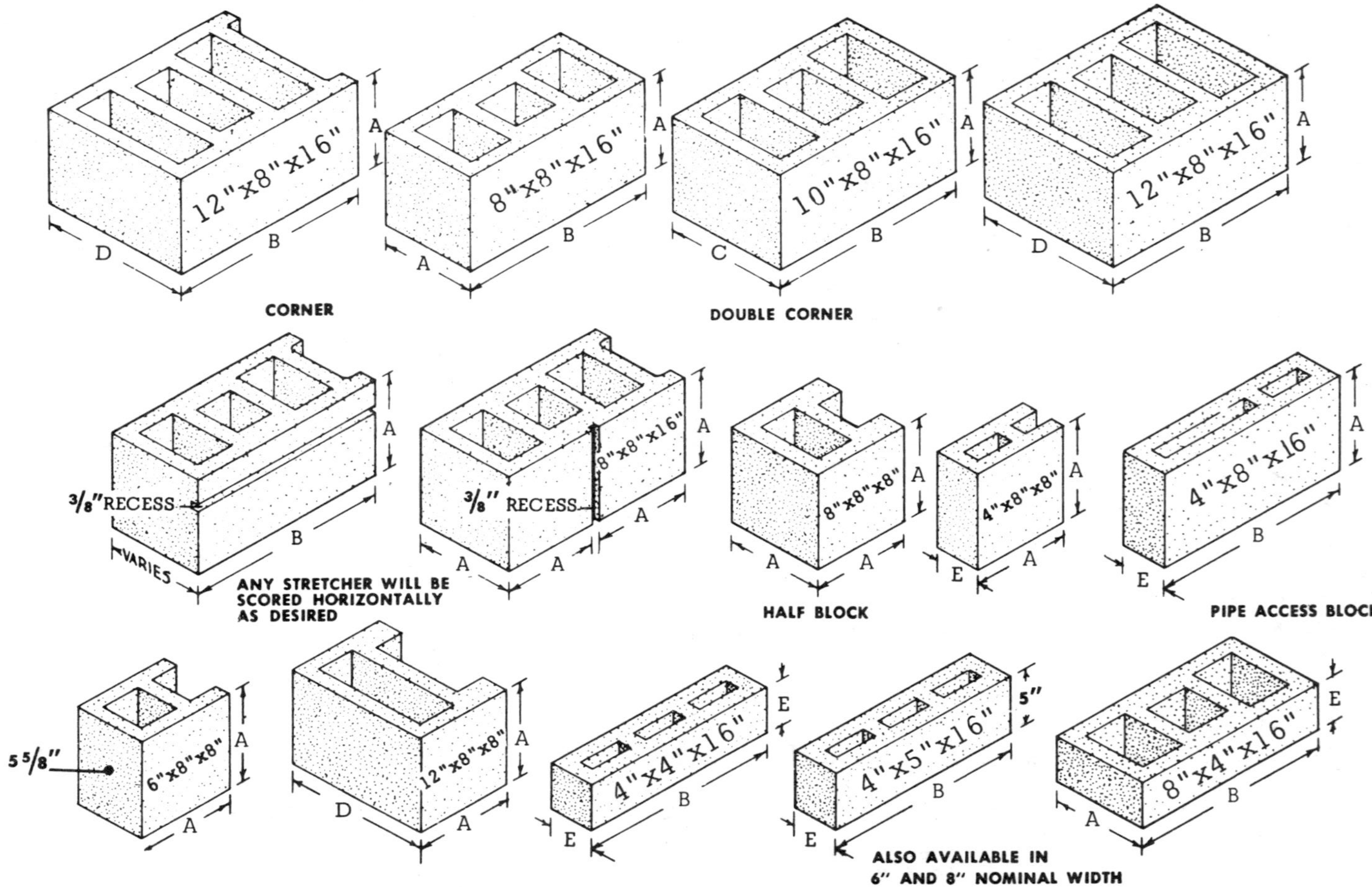
12"x8"x16"
8"x8"x16"
10"x8"x16"
12"x8"x16"
A
B
C
D
E
CORNER
DOUBLE CORNER
3/8"RECESS
VARIES
ANY STRETCHER WILL BE SCORED HORIZONTALLY AS DESIRED
3/8" RECESS
8"x8"x16"
8"x8"x8"
4"x8"x8"
HALF BLOCK
4"x8"x16"
PIPE ACCESS BLOCK
5 5/8"
6"x8"x8"
12"x8"x8"
4"x4"x16"
4"x5"x16"
5"
8"x4"x16"
ALSO AVAILABLE IN 6" AND 8" NOMINAL WIDTH

HANDY - REFERENCE - LUMBER

PLYWOOD - FLAKEBOARD - HARDBOARD - MOLDINGS

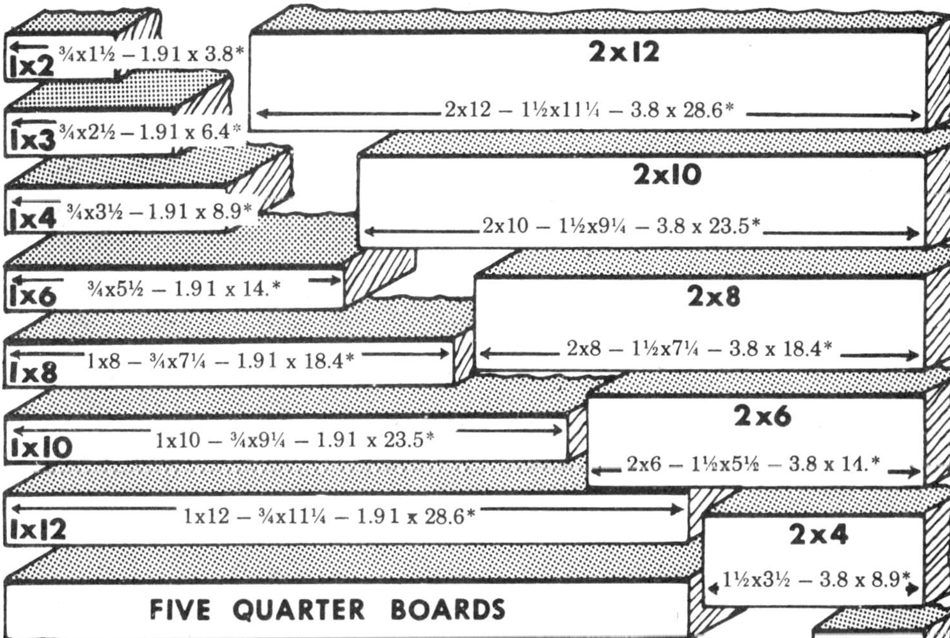

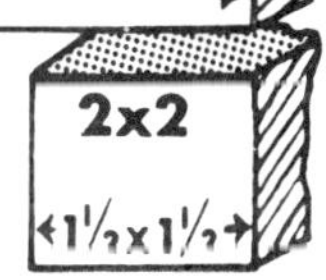

* approximate metric size

DRESSED SIZES

1" BOARDS ARE ¾" THICK
2" BOARDS ARE 1½" THICK
5/4" BOARDS ARE 1-1/16" THICK.

PLYWOOD — 4' x 8' x ¼", ⅜", ½" and ¾", interior or exterior.

FLAKEBOARD — 4' x 8' x ⅜", ½", ¾"

HARDBOARD — 4' x 6', 4' x 8' x ⅛", ¼", standard and tempered.

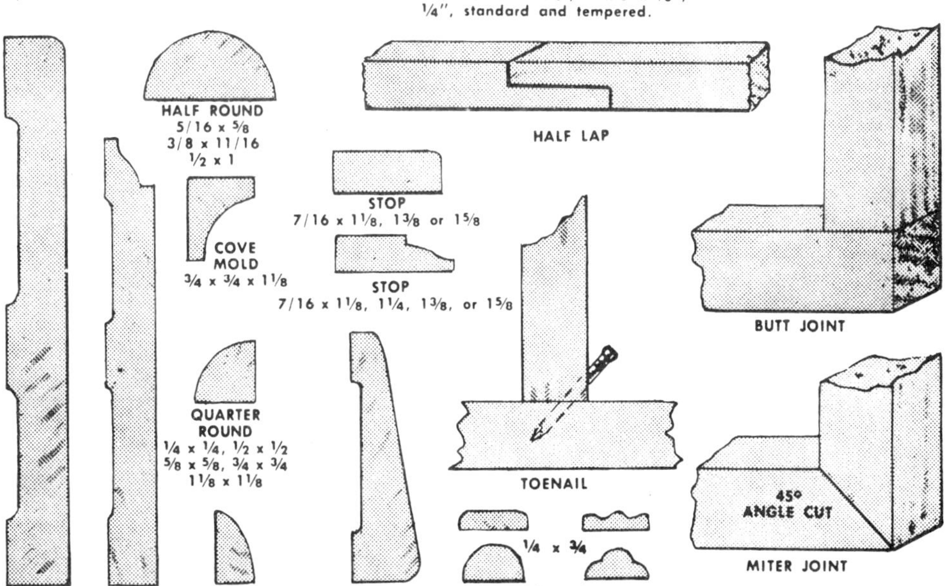

HOW TO THINK METRIC

Government officials concerned with the adoption of the metric system are quick to warn anyone from attempting to make precise conversions. One quickly accepts this advice when they begin to convert yards to meters or vice versa. Place a metric ruler alongside a foot ruler and you get the message fast.

Since a meter equals 1.09361 yards, or 39⅜"+, the decimals can drive you up a creek. The government men suggest accepting a rough, rather than exact equivalent. They recommend considering a meter in the same way you presently use a yard. A kilometer as 0.6 of a mile. A kilogram or kilo as just over two pounds. A liter, a quart, with a small extra swig.

To more fully appreciate why a rough conversion is preferable, note the 6" rule alongside the metric rule. A meter contains 100 centimeters. A centimeter contains 10 millimeters.

As an introduction to the metric system, we used a metric rule to measure standard U.S. building materials. Since a 1 x 2 measures anywhere from ¾ to 25/32 x 1½", which is typical of U.S. lumber sizes, the metric equivalents shown are only approximate.

Consider 1" equal to 2.54 centimeters; 10" = 25.4cm.

To multiply 4¼" into centimeters: 4.25 × 2.54 = 10.795 or 10.8cm.

INCH	MILLIMETER
1"	25.4
15/16	23.8
7/8	22.2
13/16	20.6
3/4	19.0
11/16	17.5
5/8	15.9
9/16	14.3
1/2	12.7
7/16	11.1
3/8	9.5
5/16	7.9
1/4	6.4
3/16	4.8
1/8	3.2
1/16	1.6

INCHES			CENTIMETERS
1			2.54
	1/8		2.9
		1/4	3.2
	3/8		3.5
		1/2	3.8
	5/8		4.1
		3/4	4.4
	7/8		4.8
2			5.1
	1/8		5.4
		1/4	5.7
	3/8		6.0
		1/2	6.4
	5/8		6.7
		3/4	7.0
	7/8		7.3
3			7.6
	1/8		7.9
		1/4	8.3
	3/8		8.6
		1/2	8.9
	5/8		9.2
		3/4	9.5
	7/8		9.8

Inches	cm
4	10.2
1/8	10.5
1/4	10.8
3/8	11.1
1/2	11.4
5/8	11.7
3/4	12.1
7/8	12.4
5	12.7
1/8	13.0
1/4	13.3
3/8	13.7
1/2	14.0
5/8	14.3
3/4	14.6
7/8	14.9
6	15.2
1/8	15.6
1/4	15.9
3/8	16.2
1/2	16.5
5/8	16.8
3/4	17.1
7/8	17.5
7	17.8
1/8	18.1
1/4	18.4
3/8	18.7
1/2	19.1
5/8	19.4
3/4	19.7
7/8	20.0
8	20.3
1/8	20.6
1/4	21.0
3/8	21.3
1/2	21.6
5/8	21.9
3/4	22.2
7/8	22.5
9	22.9
1/8	23.2
1/4	23.5
3/8	23.8
1/2	24.1
5/8	24.4
3/4	24.8
7/8	25.1
10	25.4
1/8	25.7
1/4	26.0
3/8	26.4
1/2	26.7
5/8	27.0
3/4	27.3
7/8	27.6
11	27.9
1/8	28.3
1/4	28.6
3/8	28.9
1/2	29.2
5/8	29.5
3/4	29.8
7/8	30.2
12	30.5
1/8	30.8
1/4	31.1
3/8	31.4
1/2	31.8
5/8	32.1
3/4	32.4
7/8	32.7
14	35.6
16	40.6
20	50.8
30	76.2
40	101.6
50	127.0
60	152.4
70	177.8
80	203.2
90	228.6
100	254.0

FEET = INCHES = CENTIMETERS

FEET	INCHES	CENTIMETERS
1 =	12 =	30.5
2 =	24 =	61.0
3 =	36 =	91.4
4 =	48 =	121.9
5 =	60 =	152.4
6 =	72 =	182.9
7 =	84 =	213.4
8 =	96 =	243.8
9 =	108 =	274.3
10 =	120 =	304.8
11 =	132 =	335.3
12 =	144 =	365.8
13 =	156 =	396.2
14 =	168 =	426.7
15 =	180 =	457.2
16 =	192 =	487.7
17 =	204 =	518.2
18 =	216 =	548.6
19 =	228 =	579.1
20 =	240 =	609.6

INDEX TO MONEY SAVING REPAIRS, IMPROVEMENTS, PATTERNS AND BOOKS

(Number designates Easi-Bild Pattern or Book)

Acoustical Ceiling
609,613,615,665,684,685
Acoustical Fabric 612,804
Acrylic 607,611,694,792,811
Adapter
electrical 664,694
fireplace pipe 674
plumbing pipe 675,682,685
sound system 612
Add-a-Room 609,760
Adhesives
acrylic 792
carpet 683
ceramic tile 606,816
fiber glass panel 611,631,696
floor covering 606,615,665,683,685,816
furniture 823,753,761
gluing reference guide 623
ironwork repair 617,685
paneling 605,615,665,684,685
plastic pipe 675,682
seam sealer 683
wood flooring 685
Aerators, faucet 675
Air Cooler, window 135
Air Conditioner
framing 832,665,685,773
Airplane, see Toy
Alarm devices 695
automatic telephone dialer, bell, burglar, emergency, floodlight, fire sensor, garage door, panic buttons, perimeter alarm, photo electric eye, smoke detector, ultrasonic alarm 695
under carpet mat 683,695
Alphabet, easy to trace 607
Aluminum
binder bars 683
carpet grippers 683
flashing 603,607,609,613,832,649,665, 674,680,696,751,773
foil insulation 832,685
footing scaffold 696
gutters 613,832,679,684,781
hinges 751
leader pipe 679,697
louver 832,649,665,679,697,763
moldings 605
ridge 684
roofing panels 679,680,684,696,751
screens 807
siding 684,696
skylight frame 696
sliding door track 658,758
spikes 613,679,696
square tubing 658,672,758
storm window 632
tackless strips 683
termite shield 609,832,679,697
traverse rod 627
vent 613,665
washers 751
windows 603,613,623,832,685
window casement 832
Amplifiers 612
Anchor bolts 84,609,611,613,615,617, 649,663,668,679,680,684,697
clips 611,832,679,697
Animal Cages 751
Answering Systems, telephone 695
Anti-Freeze
toilet 675,682,685
Antiquing 623,761
Anti-theft Chains 695
Apartment, garage 763
Appliance Centers 3,243,244,245,608, 832,658,685
Apron, window 605,609,679,685
Arbor, router 615,761
Armchair Repair 623
Asbestos
board 697
roofing 696
Ash Pit 674
Asphalt Roofing 609,663,685,696,763
Asphalt Tile 615,665,685,773
Astragal 611,811
Athletic Equipment 152,153,154
Attic Modernization 603,665,773
built-in lighting 694
stairways 615,665,763,773
ventilators 609,613,631,665,682,697, 763,773,781
Auger Bits 694,753,754,761
Automatic Timer 694
Auxiliary Power Generator 694
Baby Cradle 761
doll 753
match cradle 756
Backing Board 682,685
Ballcock, toilet 675
Balustrade, dollhouse 753
Bars 690
bookcase 690,804
buffet 718
cabinet 690
colonial tavern 612
corner 690
playroom 690
record cabinet 690
revolving cabinet 690
revolving shelf 508
room divider 658,758
sink 690
storage 605
table 140

TV bar 690
wall 612
Barbecue 316,668
Barge Boat 77
Barn, Red 679,680
convert into house 850
doors 679,680
tool house 679
Barnyard Pulltoys 79,756,771
Base Cabinet 3,191,243,244,245,608,658, 685,758,761
Baseboard 605,609
Basement, house 832
Basement Modernization 615
carpeting 683
drapery 605
electrical wiring 615,694
family room 615
greenhouse entry 611,615,811
luminous ceiling 615,694
outside entry 615
paneling 605,615
plyscord decorations 615
partitions 615,832
stairs 615,832,697,773
sump pump 615,617,685
tile floors 606,615,685,816
waterproofing 615,617,832,685,697
windows 615,697
Bathroom Accessories
bathtub beauty bar 327
bathtub repairs 675,685
cabinets 158,606,816
carpeting 683
fiber glass bathtub, toilet 682,685
fixtures 675,682
floors 606,816
lavatory enclosure 185,606,816
lighting fixtures 694
medicine chest repair 694
plumbing wall 682,685
sealants 606,816
sinks 675,682,685
telephone shower 682
tiles 606,816
toilet repair 675,685
vanity 606,816
ventilator 682
Bathroom Installation 682,685
bathtub, fiber glass bathtub, fittings, lavatory, medicine cabinet, plumbing tree, plumbing wall, prefab bathroom unit, septic tank, shower, toilet.
Batter Boards 609,631,832,663,679,680, 684,763,781
Battery
alarms 694,695
dollhouse lighting 694,753
floodlights 694,695,850
Beds and Bunks
colonial under eaves rope 68,761
colonial crib 761
dollhouse colonial cradle 753
dollhouse rope spring bed 753
double-deck 87,771
headboard 126
houseboat 676
studio 623,633
under bed storage 634,792
Bell Alarm 694,695,850
door 694
Benches
cobbler's 102A,586,756,761
colonial 23,586,761
dollhouse cobbler's bench 753
dollhouse fireside bench 753
dollhouse church pew 753
dining 307
fireside 23,761
lawn 325
peasant 57
work 672,677
Bi-fold Door, closet 605,634
Binders, carpet 683
Birdhouses, Feeders 669,756,807
bluebird 110,669,807
feeder 9,669,807
martin 669,807
Swedish 10
weathervane bird feeder 669,807
wren 11,111,669,756,807
Bits
auger 694,753,754,761
countersink 761
drill 613,684,761
screwdriver 761
Blackboard
kitchen message center 313,578
play table 117
Blades
saw 634,677,684,685,754,756,761,792
Blanket Storage 761,792,962
Bluebird House 110,669,807
Boards
bed 126
bulletin 93,607
fence 607,631,781,807
memo 313
Boats
barge or pram 77
cartop float 247
houseboat 676
kayak 317,757
rowboat 77,85
sail, centerboard 194
sail scooter 248
surfboard 247
toy 60,67,72,771
Body Harness
roofing 696
Bookcases 271,605,612,658,664,690, 761,804

acrylic 792
built-in 612,664,804
corner 664,804
desk 18,758,761
dollhouse 753
free standing 664,804
record 192,664,690,792,804
room divider 658,664,758,804
window 271,605,627,664,804
Bookends 756
Bookshelves
all purpose 6
applique 756
duncan 35,761
hanging 21
Bow Window
bonnet 609,668
Breaker Switches 694
Breezeway
how to build 910
how to screen 613,781
Bricklaying 668,674
barbecue 316,668
fireplace 674
mortar 668,674
patios 668
tools 668,674
tree well 668
walks 668
walls 668,685
veneering 668
Bridging 603,609,613,615,832,663,674, 679,682,684,685,697
Broom Closet 158,672
Buffet
bar 718
dining 138,718
Buildings, see Garage, Houses, Forms, Footings, Framing, Foundations, Kennel, Toolhouse, Stable.
Built-in
bookcases 612,664,804
cabinets 608,658,758,761
closet 605,615
cornice 605,627
counter range 608,658,685,758
fireplace 674
hi-fi, stereo 612
lighting 694
range 608,658,685,758
record storage 612,658,804
refrigerator 608,658,685,758
sewing center 672
sink 608,658,685,758
storage units 605,672
wall to wall 664,804
Built Up Roofing 696
Buffet Bar 718
Bulletin Board 93
outdoor 607
Bungalow, see Cabins
Bunks, see beds
Burglar Alarm Installation 695
Bushings
electrical 694
plumbing 675

Cabana 663
Cabinets
acrylic corner 792
base 3,191,243,244,608,658,685, 758,761,804
bathroom 158,606,816
broom 156,672
built-in 608,658,685,761,804
cleaning utensil 156
contemporary 658,758
corner 753,792
dollhouse corner 753
end table 141
fishing rod 630
freestanding 658,804
furniture 658,758,804
gun 266,630
hi-fi 272,612,804
kitchen 243,244,245,246,608, 658,685,758
linen 634
radial arm saw 672
record 192,436,658,690,804
sewing 543,672
stereo 612,804
storage 605,634
tool 672,677
trainboard 190,672
wall acrylic 792
wall hanging 24,608,658
wall to wall 191,192,605,608,612,658 664,719,804
wine storage 608,634
Cabins, Cottages 51,84,91,684
Cages 751
catpartment, cavy, guinea pig, hamster, kennel, parakeet, rabbit
Calking, also caulking 606,613,668, 675,682,696,697,816
Camper 594
Cane Webbing 823
Canned Goods Storage 770,608,634,758
Canopy 305,607,807
Canvas Deck 613,781
Canvas Covered Kayak 757
Cape Cod House 514
Carpeting 683
Carport 273,680
Carriage, stairs 603,615,617,832, 697,763,773
Casement Windows 609,613,781
Casing 605,609,763,773
Cast Iron Pipes 675,679,682
Cat Entry 724,751
Catpartment 751

Cavy Cage 751
Cedar Room 265,605
Ceiling Joists 603,609,613,615,832,663, 679,680,684,685
Ceiling, suspended 609,613,615,665, 684,685
Ceiling Trim Plate 674
Ceramic Tile 606,816
how to lay, repair bathroom, kitchen, patio, steps.
Chain Link Fencing 751
Chairs
child's lawn 132,754
dollhouse 753
interior 92,306
lawn 32,169,311,322R,548,754
repair, refinish, reupholster 623
Chaise 78,169,312,754
Chalk Line 613,832,668,679,684,685, 696,697
Charts, reference
alarm accessories 695
aluminum binder bars and carpet grippers 683
brackets and fasteners 623,761
carpentry tips 672
ceiling fixture and lamp repair 694
concrete block 617,832,697
drill bits 761
esssentials of brick construction 617,668
fireplace 674
glue 623
kitchen equipment template 608, 658,758
layout grid 615,650,604,605
lumber chart 605,607,609,611,612,613, 617,623,679,679,684,685,751,832
metric conversion chart 605,606,607, 608,609,617,630,631,658,664,682,685, 690,751,756,757,761,781,807,816,832
nails 672
plant guide 611,811
plumbing charts 675
beveled cone and slip joint washers, brass friction rings, cap thread, copper and cast iron fittings, dome and square stem bonnet packing, faucet stems, friction rings, lock rings, "O" rings, plastic fittings, renew seats, top bib gaskets.
roofing reference guide 696
screw chart 612,623,679
siding 696
storage units 605
wiring diagram 694
dollhouse wiring 694,753
Chests
blanket 37,761,792,962
storage 672
tool 76,672
toy and storage 37,65,962,771
wardrobe 139,658,771
Chicken House 14
Children's Furniture
blackboard message center 313,578
blackboard table 117,771
bookcases and shelves 18,21,192,664
bookends 102B,756
bulletin board 93,607
bunks and beds 68,87,761,771
desk 18,537,761
dollhouse furniture 753
giraffe clothes tree 34,771
headboard 126
lawn chair 132,754
lamp 98
magazine rack 25,920,761
picture frames 623
playtable 117,771
record and hi-fi cabinets 192,272,436, 612,804
shoe shine box 45
shoe rack 345
step stool 353,771
storage cabinets 138,605
storage chests 37,65,962,761
telephone shelf 313
toys 771
trainboard 190,672
TV tables 16,925
wall decorations 539,580,581,615,756
wall shelves 4,6,31,35,159M,756, 576,761
Child's Playhouse 148,649,811
Chimes, Swedish Door 561,672
Chimney 91,910,674,696,753
capping, constructing 674
cleaning 850
cricket 674,696
flashing 674,696
flat roof 674
dollhouse 753
housing 674
masonry 674
outside installation 763
prefabricated 832,674,685
slip section 674
China Cabinet, dollhouse 753
Christmas Displays
angel 769
angel banner 670
camels 410C
candy cane 435
carolers 331
choir boys 562
fireplace mantel 768
giant outdoor cards, murals 149,331, 540,560,562,592,769,942A,942B
illuminated displays 560,562,592,694, 767,769
indoor decorations 310,670,767,768
madonna 149,540,767

nativity
table top 310
life size 410
noel greeting 531
reindeer and sleigh 433,434
santa 431,575
window painting 149
wiring 694
wisemen greeting 560
wreath 767
Circuit Breakers 694
Cistern, how to build 617
Clamps 623,754,761
Clapboard Siding 603,609,613,832,649, 663,684,696,763
Climbing Gym 153
Climbing Pole 154
Closets 605
under stair 615,773
Closet Flanges, pipe 675,682
Clothes Tree 34,771
Clothing Storage 605
Cobbler's Bench 102A,586,756,761
dollhouse size 753
Coffee Tables, see tables
Cold Water Pipes 675,682,685
Collar Beams 679
Collars
fireplace 674
smokepipe 674
Colonial
bed 68,753,761
bench 23,753,761
blanket chest 761,962
bookcase 18,753,761
cabinet 753,761
Cape Cod house 514
candleholder 559
china cabinet 753,761
cobbler's bench 102A,586,753,756,761
corner cupboard 753
cradle rocker 753,761
cupola 609,679,680
doll cradle 753
dollhouses 753
dollhouse furniture 753
fireplace mantel 605,674,753
furniture 753,761
hutch cabinet 270,753,761
planters 102A,753,756,761,933
Portsmouth magazine stand 761
rope spring bed 823
table-bench 94,761
tavern bar 612
wall shelves 2,24,753,756,761
weathervane 524
window valance 1,27,30,157,605, 627,753
Columns
colonial dollhouse 753
concrete 631,697,781
jack post 613,615,674,685
steel 832,697
Component Systems, stereo 612,804
Concrete & Masonry 617,668,697
barbecue 316,668,674
block chart 617,832
block layout stick 617
chimney 674
colors 617
culverts, curbs 732,679,680,751
decorator blocks 582,617
floors, garage and basement, barns 273,606,613,615,617,832, 663,679,680,684,697
footprint stepping stones 96
foundations and footings 84,86,113, 609,611,613,617,832,649,663,679, 680,684,697,751
kennel run 751
mixing 617,697
patio floor 591,606,617,631,781,816
piles 567,617,631,697,781
repairs 617
steps 617,685
story pole 697
tools 617,697
tree bench and repair 617
waterproofing 615,617,663,685,697
work simplified 617
Contemporary Furniture 758
bars 508,690,718,804
bookcases 664,804
cabinet furniture 758
chairs 92,306,311,548
desk 542
dining buffet 138,718
magazine, book rack 658
planter 82
record cabinet 192,436,664,804
room divider 128,658,758,804
studio bed 823,633
tables 95,137,140,141,309,452,554
wardrobe 139,658
Cooler, Attic Air 135
Coping Saw 634,664,753,754,756,761
Copper Fittings 675,682
Corner Pole 668
Cornice
how to build 605,612
install traverse rod 605
Corrugated Ties 663,668,763
Cot, String Bed 68,823,753,761
Cottages, see Homes
Counter 80,243
sink 606,608,658,685,816
tops 608,658,685,758,816
Couplings, Pipe 675,682,685
Cove Molding
base 603,605,608,615,658,664, 665,685,753,758,804
ceiling 603,605,658,664,665,674, 753,761,758,804
Covering
pipe, girder 605,608,615

ducts 685
Cradles
baby 761
cigarette and match 756
dollhouse 753
Cricket, fireplace 674,696
Cupola, Colonial 609,679,680
Curtain Rods 605

Dado Sets 761
Damper, fireplace 674
Daybed 623,633
Deck 631,763,773,781
Decorations
garden 304,756
lawn 81,96,438,587,607,756,801
patio 582
table 17,43
wall 539,580,581,582,615,701,702,704,
705,706,709,712,713,756,761
Depth Gauges 607,677,761
Desks 18,139,537,543,658,758,761
Detectors, fire, smoke 674
Diagonal Bracing 609,613,832,679,
680,697
Dimmer Switches 694
Dining
bench 307
buffet 138,718
table 94,95,l37,554,761
Dish Storage 246,270,608,658,685,758
Displays, see Christmas Displays
case 159M,605,607,627,792
Diverter Valve 675,682
Dividers
room 128,308,608,658,758,792,804,930
outdoor privacy 582,607,617,631,668,
781,807,937
Dog Houses 28,751
Doll Carriage 59,61,753
cradle 753,761
Dollhouses 753
southern colonial, packaway,
open back colonial
Dollhouse Furniture 753
china cabinet, cobbler's bench, corner cupboard, cradle, dining and occasional chairs, drop leaf table, fireside bench, hutch cabinet, montpelier bookcase, planter, rope spring bed, shaker hutch, step stool, valance board
Door
alarm 695
bar holder 679
bi-fold 605
bumper 679
canopy 305,607,807
chimes 561,672,677
decorate with plysculpture 615,704,758
dutch door 832,649,679
furnace room door 763
garage 86,663,763
glass sliding 613,695,763,781
hayloft 679
how to install 608,609,613,615,631,832,
658,679,680,685,697,763,773
provincial trim 623
remove 608,680,684
outside entry 611,615,832,684,697,
754,763,773,781
sliding 605,634,658,679,680,754
stall 679,680
storage 605,672
toolhouse 649
trolley door hanger 679
Dormer, how to build 773
Double Hung Windows 608,609,613,684,
685,763,773,850
Downspouts, guttering 613,832,679,
684,696
Drain 675,682
Drainage Tile 615,617
Drapery
traverse track installation 605
Drawers
how to build 608,612,658,672,677,758
storage 605,634
Drills 606,615,677,682,694,695,753,
756,761,816
Drive Pins 612
Driveway
culverts, curbs 732
markers 438,587
Drop Leaf Table, dollhouse 753
Dry Well 609,615,617,832
Ducts, fireplace 674
Dutch Door 679,680

Easel, artists 555
Edging Tool
concrete 617,668
Elbows
fireplace 674
plumbing 675,682
Electric Light Gardening 611,694,811
Electrical Alarms 695
built-in lighting 694
dollhouse lighting 694,753
repairs 694
wiring 694
Emergency Power 694
End Tables 99,127,141,925
Enclosures, see Built-In, cabinets
books 612,664,804
brooms, mops 156,634,672
chimney 674
garbage can 850
hi-fi and record 192,272,436,612,804
lavatory 158,606,816
lighting 694
porch 613,781
pots and pans 578

radiator 672,677
refrigerator 608,658,758
sink 41,158
trains 190,672,677
truck body camper 594
wall oven 606,658,685,758
window, see valance 605
Entry, outdoors 611,615,617,832,684,697,754,763,773,781
Escutcheon 675,682
Excavating 84,86,113,609,611,613,617,832,649,663,679,680,684,697,751,763,781
Expansion Joint 613,617,696,697
Expansion Plug 609,613,615,631,658,781
Express Wagon 40
Exterior Siding 603,609,613,649,663,684,696,763

Fan, window 135
Farmyard Animals 79,756,771
Fascia 609,613,832,663,680,684,696,763,781
Fasteners, hollow wall 612,804
Faucet
repairs, replacement, seat stems 675
Feeders, bird 9,669,807
Fences 607,631,668,781,807
chain link 751
board 607,697,807,941
privacy 607,631,668,781,807,937
Williamsburg 315,607,807
Ferrule 675,682
Fiber Glass
bathtub 682,685
insulation 609
membrane 696
panels 611,631,696
Fire Alarms, detectors 695
Firebrick 674
Fireplace 316,668,674
apply decorative tile around 606,816
Christmas 768
fixtures 674
mantel 231,605,674,753
masonry 674
outdoor 316,668
prefabricated 674,685
radiant heat 674
Fisherman's Cabinet 266,630
Fixtures
dollhouse 753
electrical 694,695
plumbing 675,682,685
Flanges, pipe 675,682,685
Flashing
chimney 674,696
dormer 773
skylight 696
roof 609,613,832,649,674,680,696
Flexible Shower 675,682
Float, cartop swimming 247
Floating, concrete 613,617,781
Floodlights 832,694,695,850
Floor
asphalt vinyl tile 615
carpets 683
ceramic 606,816
how to lay 615,665,683,685
level 608,685,816
repair 608,685
underlay 615,683
Flue Pipe 674
Fluorescent
lighting 615,694
lights, plant growing 611,694,811
Flush Valve, toilet 675
Foam Rubber, upholstery 623
Folding
lawn bench 325
picnic table 754
screen 56
settee 55
snack table 43
table 26
Footings 84,86,113,609,611,613,617,832,649,663,679,680,684,697,751
Footprints, stepping stones 96
Form, concrete 613,617,668,697
Foundations 84,113,609,611,613,617,832,649,663,679,680,684,697,751,763,781
waterproofing 615,617,832,685,697
Frames
picture 823
window 605
Framing 832,697,763
Free Standing
bookcase 612,804
fireplace 674
room divider 658,758,804
Furniture
bars 690,804
children's 771
colonial 761
contemporary 658,758
dollhouse 753
Furniture, repair 823
antiquing 823,761
cane webbing, spring replacement, refinishing, reupholster 823
Furring 605,609,615,832,685,696,763
Fuses
wiring 694

Gable Roof, studs 609,611,832,649,663,679,680,696,697,811
Gambel Rafter Roof 679,697
Garage
carport 680
door alarm 695
doors 86,663,763
one car 680

transforming into housing 663,684,763
transform into stable 684
two car 663
two car with apartment above 763
Garbage Can Enclosure 850
Garden
greenhouse 611,615,811
electric light 611,694,811
hotbed 611,811
tool house 51,89,649,679,811
trellis 304,607,807
Garment Storage 605
Gaskets, plumbing 675,682
Gates 315,607,807,941
Gauges, depth 607,677,761
Girders, enclosing 605,615
Glass Sliding Doors 613,695,763,781
Glazing 603,609,613,623,684,685,773,781
Glider, lawn 754
Gluing, complete guide 623
furniture 623,753,761
Grade Level Establishing 609,611,613, 617,832,663,697,763
Grain Box 679
Grates, fireplace 674
Greenhouses 611,811
walk-in 611,811
window 566,811
Grout
tile 606,816
concrete 613,617,668,781
Guest House 84,684,763
Guide Lines, laying out 84,86,113,609, 613,617,832,649,663,679,680,684,697, 751,763,781
Guides, drilling 677,761
Gun
cabinets 266,630
racks 630
Gutters 609,613,832,679,684,696,763,781
Gymnasium, outdoor 152,153,154

Half Lap Joints 761
Hammer, electric jack 680,685,697
Hamster Cage 751
Handles
barn door 680
Hangers
barn door 679,680
Hardware
drapery 605
garage door 86
Harness, roofing safety 696
Hauling Cart 66,679
Headboard 126
Header, framing 832,680,697
Head, shower 675,682
Hearth
installation 674
prefabricated 674
Heat, radiant fireplace 674
Heating
hot air 685
Hi-Fi Cabinets 272,612,804
Hinges
continuous 753
door 679,680
gate 607,807
Hobby Horse 54,771
Hobby Tables, workbenches 672
Hog House 13
Hole Cutter 606,816
Hole Digger 607
Homes
convert stable into red barn house 679, 850
country cottage 91
five-bedroom Cape Cod 514
garage 663,680,684,763
guesthouse 84
hillside 103H (plans only)
low cost 832
southwest corner 432
three-bedroom modern 512,513
three-bedroom ranch 501
two-bedroom ranch 910
valley house 103V (plans only)
Home Workshop 672
Hood, hearth 674
Hot Air Circulating Fireplace 674
Hotbed, garden 611,811
Hot Plate Repair 694
Houseboat 26' 676
pontoons 600
House Numbers, signs 607,801
Houses
bird 669,807
dog 28,751
chicken 14
doll 753
duck-inn 725,751
garden tool 51,89,649,679,811
greenhouse 611,615,811
lean-to storage 89,649,751
martin 669,807
play 148,649,811
rehabilitate 685
wiring 694
Hunter's Cabinet 266,630
Hutch
cabinet 270,761
dollhouse 753
rabbit 751
Hydrant Installation 679,680
Hydraulic Jack 680,684

Increaser 675,682
Indirect Lighting 694
Insulation 609,613,632,697
Iron Pipes 682
Iron Railings, repair 617,685

Jack Hammer 680,685,697
Jack Posts 613,615,674,685
Jamb 605,697
Jeweler's Saw 753
Jig Saw Pattern Assortment 102A
  17 full size patterns 756
Jig, rafter 679,680
Jointers 617,668
Joist Hangers 609

Kayak 317,757
Kayfer Fitting 675,682
Kitchen
  blackboard 578
  broom closet 156,634,672
  buffet 138
  cabinets 243,244,245,246,605,608,634, 658,685,758
  counters 80,243,244,245,605,608, 658,685,758
  equipment templates 608,658,758
  floors 608,685
  knife rack 8
  lighting 694
  modernizing 608,658,685,758
  planning 608,832,658,685,758
  range installation 608,685,758
  refrigerator modernization 605,606, 608,685,758
  serving counter 80,243
  sink enclosure 41,158,606
  sink trap 682
  traverse track 605
  unit 3
  utility closet 156
  wall shelves 2,4,5,8
  wall oven 608,685,758
  workbench 573
Kite, Bermuda 314,771
Knick-Knack, see shelves

Ladder Tie 668
Lamps
  modern table 98
  outdoor colonial post 607,935
  planter 541
  repair 694
  rooster pin up 533
  shadow box 301,557
Lamptable 99,127,141,925
Latches
  door 680
  gate 607,807
Lavatory Enclosure 158,606,816
  P trap 675,682
  repair 675
Lawn, Patio
  bench 57,307,325,754
  chairs 32,311,322R,548,754
  chaise 78,169,312,754
  fences 607,631,668,781,807
  furniture repairs 623
  glider 754
  ornaments 9,81,96,102A,756
  settee 55,754
  tables 17,22,75,322R,554,577,754
Layout Stick
  bricklaying 668,674
  concrete 617,697
  clapboard siding 609,763,781
  tile 606,816
Lazy Susan 608,672,677,758
Lead Sealants, pipe 682
Lean-To Tool Shed 89,649
  kennel 751
Letter Boxes 607,807
Level-Transit 668,697,763
Lighting 694
  built-in cornice, valance 605,694
  dark corner 557
  luminous ceiling 615,694
  outdoor 832,694,695,763
  wall sconce, candle 559
  walls, soffits, cove 694
Line Holder 668
Lintel, fireplace 674
Lock, alarm 695
Lounge, chaise 78,169,312,754
Louver
  how to install 832,665,679,763,773
Lumber Chart 607,672,754,832

Magazine Rack 25,658,761,920
Magnetic Catch 605,758
Magnetic Contact Alarms 695
Mailbox, roadside 607,807
Mailbox Gate 607,807
Mantel
  fireplace 231,605,674,753
  dollhouse 753
Martin House 669,807
Masonry, cement 668
Masonry Drill 615,680
Masonry Fireplace 674
Medicine Chest, repair 694
Message Center, blackboard 313,578
Metal Clips 668
Metal Repairs, roofing 696
Meter, multitester 694
Modern Furniture 658,758
  see furniture, bookcase, table
Modernization
  attic 773
  basement 605,615
  bathroom 606,682,685,816
  kitchen 608,658,685,758
  lighting 694
  metal and wood cabinets 605,608, 685,758

refrigerator 605,608,685,758
Mortar, how to mix 617,668,685
apply 609,613,617,668,685
Mortising 607,664,761
Music Wall 612,804

Nailing Machine 605
Name Plates, signs 438,587,607,792, 801,807
Night Tables 343

Oriental Plysculpture 704,712
Ornamental Ironwork 617
Ornaments, see Decorations
see Christmas 694
door chimes 561,672,677
driveway 438
lawn 9,81,96,102A,617,631,756
sign & nameplate 607,756,807
table top 46
wall hanging 301,539,559
Outbuildings, see Houses
Outdoor Furniture, how to build
bench, chair, chaise, glider, table, picnic table 754
Outdoor Lighting 694
Outlets, electrical 694
Ovens, built-in 608,658,685,758
Overhead Garage Door 86,663,763

Panels
aluminum roofing 679,680,684, 696,751
ceiling 609,613,615,665,684,685, 763,773
circuit breaker 694,695
fiber glass 611,631,696
plexiglas 792
polycarbonate 811,850
Paneling, how to apply 605
Paneling Arches 605
Parakeet Cage 751
Parging, also pargeting 617,668
Parquet, flooring 685
Partitions 609,613,615,832,658,758
dollhouse 753
stable 679,680
Patio
decorations, see ornaments
furniture, see lawn
how to build 606,617,631,668,781,816
how to screen 631,781
lighting 694
paver tile 606,617,631,781,816
Peasant Bench 57
shelf 2,4
Perimeter Alarm System 695
Pet Projects 751
cat shelter, cavy cage, dog house, kennel and run, duck house, hamster, parakeet cage, rabbit hutch
Photo Electric System 695
Picnic
benches 22,57,325,754
tables 22,554,577,754
Picture Frames 823
Pier, concrete 607,609,668,763
Pilaster 664,668
Pin-up Lamp 533
Pipe Rack 49,630
Pipes
cast iron 675,682
chimney 674
drain 611,675,679,680,682
enclosure 605,615
galvanized 611,682
kitchen 675,682
lavatory 606,675,682,811
plastic 675,679,682
repair 675
smokepipe 674
Pipe Scaffold 668,674,763
Planters
box 807
dollhouse 753
indoor 46,82,102A,756,761,933
lamp 541
Plant Box Rack 611,811
Plaques, child's wall 102A,756
Plastic Laminate
how to apply 608,658,685,758
Plastic Panels 611,631,696,811
Platforms
raised platform for hearth 674
stairway 603,615,617,832,665,773
Playground Equipment 20,63,152,153,154
Playhouses 148,649,811
Playroom Bar 612,690
Playroom Furniture
child's blackboard playtable 117,771
giraffe clothes tree 34,771
tot's step stool & TV chair 353,771
toy chest on wheels 65,771
Plumb, how to 609,611,613,615
Plumbing
bathtub installation 682
cast iron fittings 675,682
charts, closet bends, continuous waste, copper fitting, diverter valve, drain, drum trap, faucet, ferrule, fiber glass bathtub, fittings, fixtures, framing, fresh air inlet, glossary, gypsum board application, horizontal branch increaser, kayfer fittings, kitchen faucet replacement, kitchen sink trap, lavatory, lavatory faucet replacement, lavatory installation, lock ring, O-ring, pipes, plastic pipe, plumbing wall, repair, replace toilet bowl seal, revent, rough in, septic tank, slip joint, soil pipe, solder cup end, spigot, stack, stack

vent, sump pump installation, telephone shower, wall frame, water pipe riser, water tank repair, toilet 675, 682,685
Plunger, toilet 675,682,685
Plywood, how to panel 605
Polycarbonate 811,850
Polyethylene 609,613,617,663,811
Pontoon 600,676
Pony
book ends 102A,756
rocker 53,771
Porch
add-on 567,631,781
build, columns, enclose, screening 613,631,781
sliding glass doors 613,763,781
Post, fench 607,781,807
Post Driver 751
Post Hole Digger 607
Potholders, holder 756
Power Tool Workbench 672
Pram 674
boat 77
Privacy Partition 607,631,668,781,937
Propane Torch 617
Pumps, sump 615,617,675,685

Quarry Tile 606,617,816

Rabbit Hutch 751
Racks
can storage 634,672,758,770
cassette tape 612,804
clothes 605
gun 630
magazine 25,758,761,920
pipe, smoking 49,630
shoe 345
storage 605,634
tool 672
towel 29
Radar Alarm 695
Radial Arm Saw
workbench 672
Radiator Enclosure 672
Radio and Stereo
bar 690
cabinets 272,612,804
how to build-in 612
Rafters 609,611,631,832,663,679,680,684, 685,696,697,763,773,781
Rail Fencing 607,781,807,941
Railings, iron 617,685
Rain Cap, chimney 674
Range Installation 608,658,685,758
Razor Saw 753
Record Cabinet 192,436,658,664,690,804
Record Changer, recorders, tapes, 612
Recreation Room 615
furniture, bar 612,690
how to modernize a basement 615
lighting 694
Rectangular Tie 668
Reducers, pipe 675,682
Refinishing Furniture 823,761
Refrigerator
enclosure 608,658,685,758
modernizing 605,608,685,758
Reinforced Masonry 668
Reinforcing Rods, wire 617,668,697
Relax-a-Board 161
Relief Valves, plumbing 675,682
Remodeling 609,685,781
Remote Detector Alarm 695
switch 695
Repair
basement walls 615,617,685
cane webbing 623
concrete, masonry 617,668
electrical 694
furniture 823
picture frame 823
plumbing 675
roofing 696
tile 606,816
window sash 685
window screening 807,850
Retaining Walls 668
Reupholstery 823
Revolving Tool Turret 672
Ridge Caps
roofing 696
shingles 696
Rifle Cabinets and Racks 630
Rings
drapery 605
plumbing 675
Rip Saws 613,664,677,754,761
Roadside Mailbox Rack 607,807
Roadside Sign 607,792,807
Rocker
colonial crib 761
dollhouse cradle 753
pony ride 53,771
Rod
reinforcing 617,668,697
ties 697
traverse 627
Roll Roofing 696
Roller
barn door 679,680
garage door 86,680,684
Roofing Repairs and Application 696
asphalt, built-up, metal, wood, slate, asphalt shingles, roll roofing
chimney cricket 674,696
dollhouse 753
flashing 603,609,696
heating cable, tape 694

metal 696
safety harness, scaffold 696
vent 674,675,679,696
Room, add on 609,760,773
attic 773
basement 615
bathroom 609,682,685,773
dividers 128,308,608,658,758,930
Rope, safety harness 696
Routers 608,697,753,758,761
Rowboats 77,85
Rug 683
adhesive, tape, backing, binder bars, grippers, padding, seaming, under carpet alarm 683,695
Rush Weaving 623

Saber Saw 674,677,685,753,754,756,761
Saber Saw Patterns 102A,756
Safety
alarms 695
harness 696
relief valves 675,682,685
signs 438,587
switches, electric 694
Sailboat 194,248
Sandbox 20,77
Sanitary Tee 682
Sash, window
alarm 695
cellar 663,679
repair 685
steel 603,609,613
Sawhorse Toolchest 672
Sconce, candle wall 559
Screed 617,631,663
Screen Enclosure 613,631,781
Screen, folding 706,712
Screens, replacing wire 807
Screw Chart 612,623
Scribing 605,609
Scroll Saw Patterns 756
Sealing Cord 631,781
Seaming Tape, carpet 683
Sectional Bookcase 664,804
Security Devices 695,850
auto, fire, door, garage door, fire sensors, under carpet, window
Service Center 243
counter 80
Settee, lawn 55,754
Sewer Line 675,682,685
Sewer Rod 675
Sewing Cabinet 634,672
table 543, 823
Shadow Box 301,557
Shaker Hutch Table 94,761
dollhouse furniture 753
Sheathing, exterior 603,609,613,832
Shelf Standards 664,792
Shelves, 2,4,5,6,8,21,24,605,612,658,672
Shingles, asphalt 113,696,910
roofing 432,609,663,696
Shoe Rack 345
Shoe Shine Box 45
Showers and Fittings 675,682,685
Shutoff Valves, plumbing 675,682,685
Shutters, dollhouse 753
Siding 603,609,613,832,663,684,696,763
dollhouse 753
Signs, nameplates 438,587,607,801,807
Sill Beam 609,832,684,697
Single Handle Faucet 675
Sink
bar 608,658,690
bases 158,606,608,675,682
counter installation 606,608,658, 685,758,816
enclosure 41,158
faucets, fittings 608,675,682,685
Skylight, how to install 696,773
Slab Construction 680
Slant Board 161
Slate Roofing 696
Slide, child's 63
Sliding Door, wardrobes 139,658
Sliding Stable Door 679,680
Smoke Detector 695
Smokepipe, adapter elbow 674
Snake, sewer 675
Snips, aviation 674
Soffit 603,605
Soil Pipe 682
Solder Ring 695
Sound Amplifier 612
Spacers, tile 816
Speakers
intercom 695
radio, stereo, tape 612,804
Speakers 612,804
Spindle 675
Sportsman Revolving Cabinet 630
Springs, furniture, cord bed 623,761
retying, replacing, no-sag 623
Stable 679,680
convert into Red Barn House 850
Stacking Tables 43
Stair Carriage 615,832,697,763,773
Stairs, concrete 613,615,617,631,685,781
concrete repair 685
how to build 615,617,832,763,773,781
Stall Door 679,680
Stands
drill 677
mailbox 607,807
radial arm saw 677
television 925
Star Drill 679,680
Starter Tee 674
Steel Girder Bridging 832
Steps 603,613,615,617,832,685
Step Flashing 603,609,696,773

Stereo Wall 612,804
Stilts 552,771
Stool, sill 605,609,613,832
Stool, tot's step 353,771
Storage 605
  bottles 758
  cabinets, see cabinets 672
  canned goods 608,758,770
  chest 37,65,672,761,962
  doll 159M,605,679,792
  garden tool house, lean-to 8,649,679
  headboard 126
  room divider 128,308,608,658,758
  sewing 543,634,672
  tool 672
  trainboard 672
  underbed 37
  undersink 41,158,606,816
  understair 605,615,773
  walls 190,605,612,658,758
  wine storage 690
Storm Collar 674
Storm Windows 303,792
Story Pole 668,674
Stringer, stair 603,615,832,665,697
  dollhouse 753
Structural Lighting 694
Studio Bed, string 623,633,761
  dollhouse 753
Studs 84,91,501,502,609,697
Subflooring 603,606,832,763,773,816
Sump Pump 615,617,675,685
Sundeck 631,763,773,781
Sunhouse 611,811
Surfboard or Swimming Float 247
Suspended Ceiling 609,613,615
Swimming Pool Enclosure 631,781
Swings 152,754
Switch Lock 695
Switch, wall 605,694

"T" Hinge 679,680
Tables
  bar 140,690
  bench 672
  bridge 95
  child's 117
  coffee 52,140,309,326,452
  colonial 94,761
  dining 94,95,137,554
  end 99,127,141
  folding 94,761
  lawn or terrace 17,75,326,554
  picnic 17,22,577
  roundtop 75,326
  serving, see coffee
  sewing 543
  TV 16,923
  wall, attached 26,774
  workshop 568,573,672
Tape Recorder, built in 612
Teenuts 672
Telephone
  dialer alarm 695
  shower 675
Termite Proofing 850
Termite Shield 609,832,697,763
Terrace 631,668,781
  table 75,326
Tile, see Ceramic 606,816
  asphalt 615
  counter 606,816
  how to lay 606,615,816
  patio design 617
  quarry 816
  vinyl 615
Toilet
  installation 682
  repairs, replacement 675
Tool Chest 71,672
Tool Houses
  garage 113,680
  garden 649,811
  lean-to 89,649
  red barn 679
Tote Box 677
Tourist House 84,684
Towel Rack 29
Toys
  Airplane 70
  Animals 79,83,771
  artists easel 555
  boats 60,67,771
  carriage 59,771
  chest 771
  circus group 48,771
  climbing, see playground
  clown 47,771
  dollhouses 753
  furniture, see Children's
  glider 70
  gym, see playground
  hobby horse 54,771
  machine gun 62
  playground equipment
    climbing pole 154
    complete gym 152
    monkey bar 153
    sandbox 20,77
    slide 63
  playhouse 148,649,811
  playtable 117,771
  pony rocker 53,771
  stable 679
  step stool 353,771
  stilts 552,771
  storage 65,605,771
  toychest 65,771
  wagon 40
  wheelbarrow 66
  workbench 672

Trainboard 190,672,677
Transformer 694,695,753
Trap 675,682
bathtub, drum, lavatory, sink, waste line
Traverse Track 605
Tree bench 617
Tree Box Planter 807
Tree Well 668
Trellis 607,631,807
Trim
apply 615
charts 623,664
Trolley Door Hanger, Rail, Bracket 679
Trophy Cabinet 630,792
Truss Rafter 697
Tub Sets 606,682,685,816
Tungsten Carbide Tip Drills 615,674
Turntable 612,804
TV Table 16
bar 690
Two Handle Faucets 675

Ultrasonic Alarm 695
Under Bed Storage 761,792
Under Stair Storage 615,697
Upholstery, repair 623
Utility Closet 156,634

Valance
bookcase 271
indirect lighting 157,694,804
window 1,27,30,550,627
Valley Flashing 696
Vanity 658
bathroom 606,816
Veneer, brick 668

Wall
cabinets 658,758
framing 663,684,697,763
gypsum board 605,763,773
oven 608,658,685,758
paneling 605
plaque, child's 102A
remove 608,685,758
retaining 668
shelves, see shelves
storage 605
switch 605,694
walks, brick 668
waterproof 615,617
wiring 694
Wall Decorations 97,159M,539,580, 581,702, see ornaments
Wardrobe 139,193,263,605,658
Washers
faucet 675
Water Pipe Riser 682
Water Shut Off valve 675
Waterproofing 617
cement 617,832,697
Wattage Chart 694,763
Weathervane 524,588
bird feeder 669,807
Webbing, furniture 623
Wheelbarrow 66
Windbraces 832,680
Window
bookcase 271
bow, picture, installation 609
framing 159
glazing 613,623,685,773,781
greenhouse 566,611,811
guards 679
how to install 609,613,679,781
how to move 608
how to repair window 685,773
screen wire replacement 807
storm 303,792
valance 1,27,30,550,605
Wire Clips 623,773,685
Wire Hangers 617
Wire Ties 685
Wireless Alarm 695
Wireless Intercom unit 695
Wiring, electric 694
Wood Conversion Chart 613
Wood Shingles 696
Workbench 568,672
Workshop 672
Worktable 573,672
Wrecking Bar 684,697
Wren House 11,111,669,756,807

"Z" Tie 668

EASI-BILD LEARN TO EARN BOOKS

#850 HOW TO FIND A JOB, START A BUSINESS

Of special interest to teens, retirees and anyone who wants to earn extra income. Learn to offer what others want to buy. No capital investment required. 210 pp., 304 illus.

#609 HOW TO BUILD AN ADDITION

Creating additional living space can prove to be one of today's soundest investments. Step-by-step directions explain how to build a 12 x 16', 16 x 24' or any other size one or two story addition, with or without an outside entry. 162pp., 211 illus., simplify every step.

#615 HOW TO MODERNIZE A BASEMENT

Whether you create a family room or turn a basement into an income producing one bedroom apartment with an outside entrance, you will find all the information needed. It explains how to install an outside entry, build stairs, frame partitions, panel walls, lay floor tile and much more. 98pp., 135 illus.

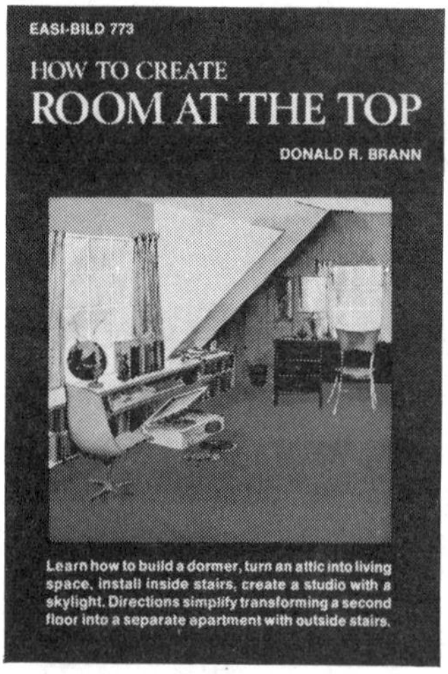

#773 HOW TO CREATE ROOM AT THE TOP

Transform an attic into extra bedrooms or an income producing apartment. Learn how to build a dormer, install a skylight, build inside and outside stairs. Creating living space provides a key to financial survival. This book tells ALL. 162pp. 239 illus.

#617 CONCRETE WORK SIMPLIFIED

This book explains everything you need to know to mix concrete, floating, finishing, grooving, edging and pointing, to setting ironwork and anchor bolts. It also explains how to waterproof a basement, install a sump pump, an outside entry and make all kinds of concrete repairs. 194pp., 257 illus.

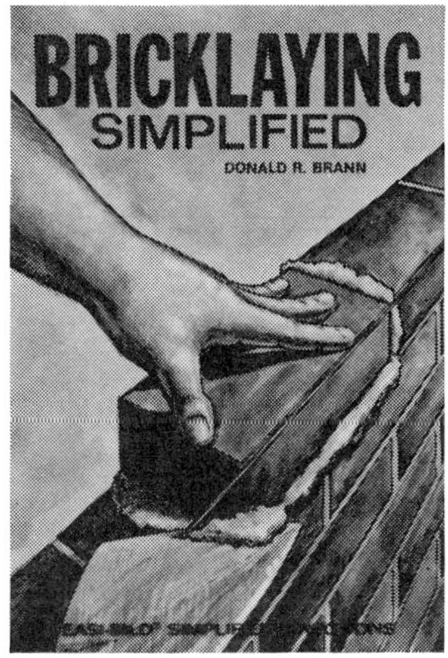

#668 BRICKLAYING SIMPLIFIED

All who seek income, peace of mind, an economical solution to a costly problem or employment in a trade where opportunity is unlimited, find this book a real guide to better living. It explains how to lay bricks, a wall, walk, veneer a house, build a barbecue, etc. It turns amateurs into pros. 146pp., 212 illus.

#696 ROOFING SIMPLIFIED

This "business of your own" book turns amateurs into professional roofers. Learn to repair or replace an asphalt, wood or slate roof; apply roll roofing, make a roofer's safety harness, walk and work on a roof with no fear of falling, plus much more. 130pp., 168 illus.

#697 FORMS, FOOTINGS, FOUNDATIONS, FRAMING, STAIR BUILDING

This book tells every reader how to get into the building industry. Whether you build your own house, buy a prefab or want a career in building, this book tells everything you need to know about forms, footings, foundations, framing and stair building. 210pp., 310 illus.

#758 HOW TO MODERNIZE A KITCHEN
Of special interest to homeowners who appreciate the convenience and Capital Gains of a completely modernized kitchen. Directions explain how to build base and wall cabinets to fit any space available or modernize existing wood or metal cabinets. Plan the kitchen that fills your family needs today and in the future. 210pp., 263 illus.

#674 HOW TO INSTALL A FIREPLACE
Everyone who wants to install a woodburning stove, build a brick fireplace or install a prefabricated metal fireplace and chimney, will find all the direction they need. Installing a chimney completely within or recessed flush with an outside wall is clearly explained and illustrated. 242pp., 354 illus.

#683 CARPETING SIMPLIFIED
Laying carpet in your home can provide the experience needed to do the same work for others. Step-by-step directions explain how to install every type of carpeting over any kind of floor, with or without padding. Learn how to carpet stairs, install under-carpet electronic alarm mats and more. 178pp., 223 illus.

#695 HOW TO INSTALL PROTECTIVE ALARM DEVICES
Recapture peace of mind by protecting all doors and windows with professional devices. Discourage a break-in with magnetic contacts that automatically notify policy, install alarm bells, detect movement instantly with easy to install radar. A layman's guide to professional installation of alarm devices. 130pp.,146 illus.

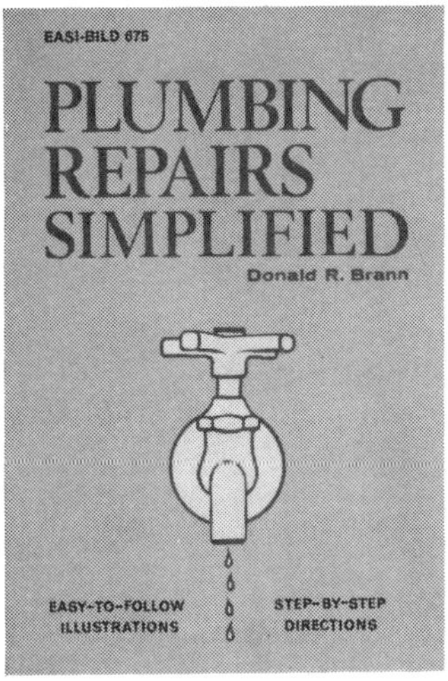

#675 PLUMBING REPAIRS SIMPLIFIED
Homeowners who dislike having their budget and peace of mind destroyed by a faulty plumbing fixture find this book helps save time, temper and money. Everyone who has learned to bake a cake or drive a car can easily replace parts and make repairs like a pro. 194pp., 820 illus.

#682 HOW TO ADD AN EXTRA BATHROOM
This complete, easy to read guide to home plumbing helps make a dream come true for only the cost of fixtures. Directions explain how to make the installation and save a bundle. Those who don't want to do any plumbing still save by preparing the area, then having a plumber make the installation. 162pp., 200illus.

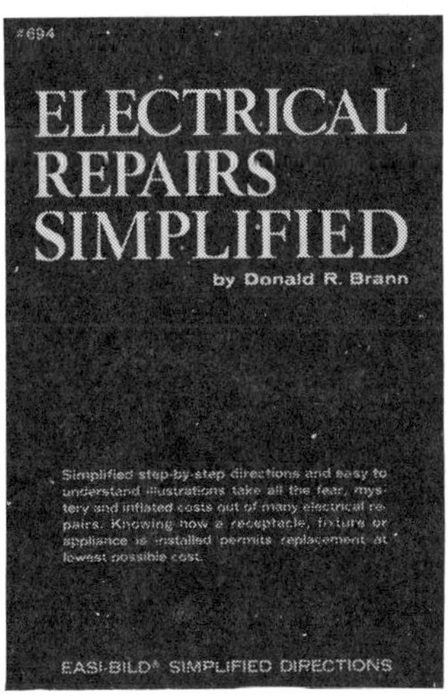

#694 ELECTRICAL REPAIRS SIMPLIFIED
Learning to economically make electrical repairs not only generates peace of mind, but also income in your spare time. This book takes the fear, mystery and inflated cost out of many troublesome repairs. A special feature explains how to install wiring in a dollhouse. 134pp., 218 illus.

#816 HOW TO LAY CERAMIC TILE
Learn how to apply ceramic and quarry tile to floor, walls, countertops and patios. Information about tools, materials, surface preparation, special situations and valuable time saving trade tips included. Directions explain how to build a quarry tile pad for woodburning stoves, how to apply decorative tile around a fireplace and more. 178pp., 225 illus.

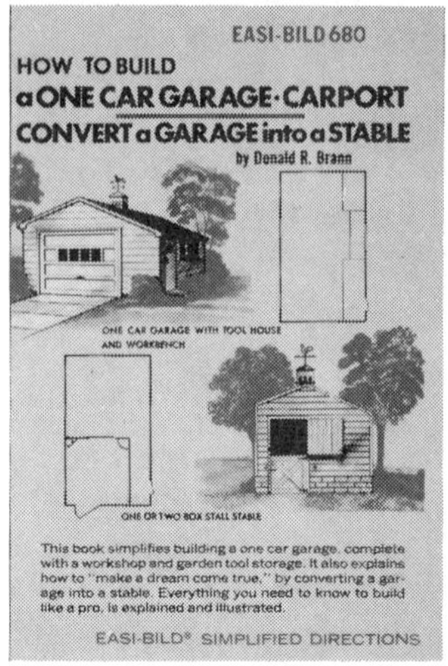

#663 HOW TO BUILD A TWO CAR GARAGE, LEAN-TO PORCH, CABANA

Building a garage can prove to be a richly rewarding experience. Letters from readers who built this garage confirm the task altered their outlook on life. Many who build turn it into an income producing singles apartment. 130pp., 142 illus.

#680 HOW TO BUILD A ONE CAR GARAGE, CARPORT, CONVERT A GARAGE INTO A STABLE

Building a one car garage with ample space for a workshop, or turning a one car garage into a two box stall stable is clearly explained. Directions tell how to raise a garage to obtain needed headroom, build a carport, lean-to toolhouse and a cupola. 146pp., 181 illus.

#684 HOW TO TRANSFORM A GARAGE INTO LIVING SPACE

Transforming a garage into a living-bedroom, with a kitchen and bathroom, can provide a safe and economical solution to a costly nursing home problem. It can also become an important income producer. Step-by-step directions explain every step. 130pp.,139 illus.

#763 HOW TO BUILD A TWO CAR GARAGE WITH APARTMENT ABOVE

Build a two car, two story garage or add a second story apartment to an existing garage. Space above provides a living, bedroom, kitchen and bathroom. Ideal for a single or couple. 194pp., 226 illus.

#754 HOW TO BUILD OUTDOOR FURNITURE
Easy to follow step-by-step directions, plus a big foldout full size pattern, simplify tracing and cutting all parts to exact shape required. Learn how to build curved back lawn chairs, a matching settee, four passenger lawn glider, a chaise on wheels and much, much more. 130pp., 174 illus, plus full size pattern.

#781 HOW TO BUILD A PATIO, PORCH AND SUNDECK
Simplified directions take all the inflated cost out of building a front or back porch, a patio to length and width specified or to size desired, a carport and sundeck. Every step, from laying footings to installation of railings, is illustrated. Make screens, porch repairs, swimming pool enclosure and much more. 146pp., 220 illus.

#807 HOW TO BUILD OUTDOOR PROJECTS
Directions explain how to make over 20 outdoor projects. Learn to repair and replace window and door screening, build gates and fencing, door canopy, trellis, outdoor planter, signs, birdhouses, feeders and more. 210 pp., 212 illus.

#811 HOW TO BUILD GREENHOUSES — WALK-IN, WINDOW, SUNHOUSE, GARDEN TOOL HOUSE
Of special interest to everyone who enjoys the fun and relaxation of growing plants the year round. The sunhouse appeals to sun lovers who enjoy winter sunbathing. 210pp., 229 illus.

#753 HOW TO BUILD DOLLHOUSES & FURNITURE

To create a memory a little girl will never forget, build one of the three dollhouses offered in this book. Those searching for a part or full time money making hobby find a ready market for dollhouses. Full size patterns simplify making fourteen pieces of dollhouse furniture. 194pp., 316 illus.

#757 HOW TO BUILD A KAYAK

Simplified directions and full size frame patterns permit building this extremely light yet sturdy kayak to three different lengths, 14'3", 16'9", or 18'0". It can easily be carried on a cartop rack and used by one or two adults. Patterns insure cutting each frame to exact size required. This book includes full size patterns for all frames.

#771 TOYMAKING AND CHILDREN'S FURNITURE SIMPLIFIED

As every reader soon discovers, toymaking possesses a certain magic. Directions for pony ride rocker, giraffe clothes tree, bunk beds, toy chest and 12 other projects included. 226pp., 330 illus., plus a full size foldout pattern.

#792 HOW TO BUILD COLLECTORS' DISPLAY CASES

Learn to build handsome, clear acrylic, museum quality, floor, table top and wall display cabinets. These provide the perfect way to display every kind of possession from dolls, china, figurines, etc. Retailers buy these cases for store use as readily as for resale. 194pp., 229 illus.

#832 HOW TO BUILD A LOW COST HOUSE — ABOVE OR BELOW GROUND

A one story house over a full basement provides alot of easy to own living space. An even more economical house to build, heat and cool is the "earth sheltered foundation house" where only the roof and front are exposed. 226pp. 117 illus.

#669 HOW TO BUILD BIRDHOUSES AND BIRD FEEDERS

Encouraging a child to build feeders and bird-houses can stimulate a lifetime interest in woodworking. Full size patterns not only simplify building but also insure success. Helping a child turn a piece of wood into a useable and saleable article builds instant self confidence. 66pp., 86 illus.

#679 HOW TO BUILD A STABLE AND RED BARN TOOL HOUSE

A 20x30' three box stall stable is easy to build. Every step of construction, from selecting a site to building the cupola, is explained and illustrated. Directions also simplify building an 8x10' or larger tool house. 178pp., 197 illus.

#751 HOW TO BUILD PET HOUSING

Encourage all who love pets to build the shelter each needs. Learn how to build a doghouse, lean-to kennel, rabbit hutch, duck-inn, parakeet cage, an all weather cat entry, plus a unique catpartment that's easy to sell, easy to rent. 178pp., 252 illus.

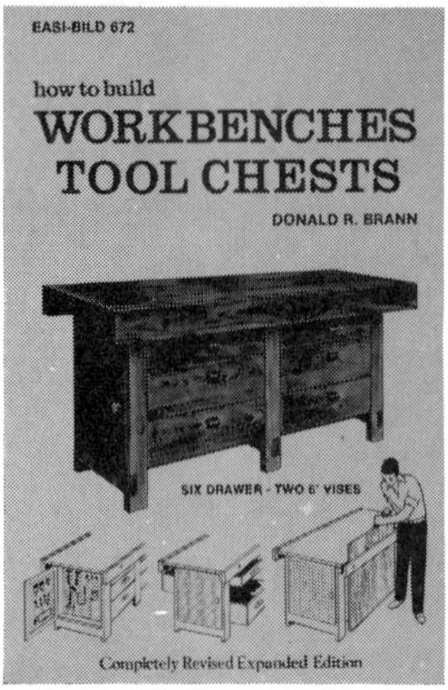

#672 HOW TO BUILD WORKBENCHES
Every home, apartment and place of business needs a workbench to economically make repairs and improvements. Build a 6' workbench with a 6' vise on one or both sides, big drawers and tool compartments. Directions for foldaway wall benches that require only a minimum of floor space also included. 180pp., 250 illus., plus full size foldout pattern.

#823 HOW TO REPAIR, REFINISH, REUPHOLSTER
Learn to apply first aid to ailing furniture. Reglue joints, replace tenons, torn webbing, springs, etc. Make a 28'' picture frame clamp, a professional mat cutting board, a rope spring bed, and a 114'' quilting frame. Everything you need to kow from tacks to tools. 176pp., 215illus.

#605 HOW TO INSTALL PANELING
Learn to apply paneling like a pro. Building a matching wall to wall storage closet with sliding doors, a fireplace mantel, install valances with indirect lighting, even build a cedar lined storage room. 146pp., 214 illus., plus full size valance patterns simplify every step.

#685 HOW TO REMODEL BUILDINGS
With abandoned big city housing units available to all who are willing to rehabilitate and occupy same, this book explains how to become a landlord with an investment of time and effort. It tells how to turn an abandoned multi-family building, store, garage or warehouse into rentable housing. Every step explained and illustrated. 258pp.,345 illus.

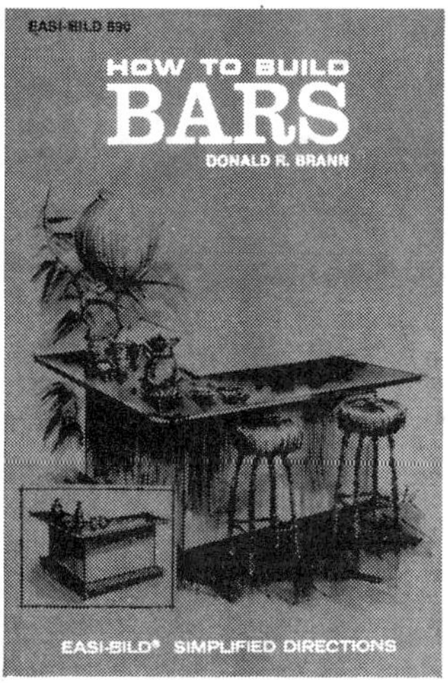

#690 HOW TO BUILD BARS
Building a bar offers a fun way to furnish a recreation room. Learning to build a straight, L-shaped or any of the seven bars described provides an easy way to start a part or full time business. Doing something today you didn't know how to do yesterday broadens one's sphere of activity. 162pp., 195 illus.

#804 HOW TO BUILD BOOKCASES AND STEREO CABINETS
Takes all the mystery and ⅔ the cost out of building bookcases and cabinets to fill any space available. Directions explain how to build wall-to-wall, built-in, free standing and sectional bookcases and stereo cabinets. 194pp., 232 illus.

#630 HOW TO BUILD SPORTSMAN'S REVOLVING STORAGE CABINET
Directions simplify building a glass enclosed gun cabinet, wall racks and a 24 x 72" revolving cabinet that stores everything from guns to clothing. Learn to make what others want to buy. 98pp., 121 illus.

#761 HOW TO BUILD COLONIAL FURNITURE
You can easily obtain furniture at a fraction of retail cost by building colonial reproductions. Easy to follow directions and full size patterns simplify building a cobbler's bench, hutch cabinet, blanket chest, under the eaves rope bed, wall cabinet and more. 258pp., 342 illus.

#756 SCROLL SAW PROJECTS
Helping everyone, a child or retiree, successfully turn a piece of wood into a handsome, useable and saleable article, builds the ego. This book insures success. 27 full size patterns permit tracing all parts, then assembling each in exact position shown on pattern. 130pp., 146 illus.

#600 COMPLETE EASI-BILD CATALOG
Anybody can do anything if they follow directions offered in Easi-Bild Books and Full Size Patterns. The catalog illustrates hundreds of patterns and home repair and improvement books. Give this book to a youth searching to find a career and many will soon be building everything from furniture, boats, garages to houses. Getting experience in their own home encourages doing the same work for others.

LIMITED EDITIONS STILL AVAILABLE

#603 How to Build a Dormer—82pp.,114 illus.
#606 How to Lay Ceramic Tile—98pp.,137 illus.
#607 How to Build Fences,Gates,Outdoor Projects—162pp.,212 illus.
#608 How to Modernize a Kitchen—82pp.,118 illus.
#611 How to Build Greenhouses—114pp.,110 illus.
#612 How to Build Wall-to-Wall Cabinets, Stereo Installation Simplified—130pp.,165 illus.
#613 How to Build or Enclose a Porch—82pp.,112 illus.
#623 How to Repair,Refinish and Reupholster Furniture — 98pp.,138illus.
#627 How to Make Cornice Boards, Draperies, Valances, Install Traverse Track—66pp.,117 illus.
#631 How to Build Patios and Sundecks—98pp.,133 illus.
#634 How to Build Storage Units—98pp.,145 illus.
#649 How to Build a Garden Toolhouse,Child's Playhouse—82pp.,107 illus.
#658 How to Build Kitchen Cabinets, Room Dividers and Cabinet Furniture—98pp.,134 illus.
#664 How to Construct Built-In and Sectional Bookcases—98pp.,137 illus.
#665 How to Modernize an Attic—82pp.,86 illus.
#677 How to Plan and Build a Home Workshop—98pp.,133 illus.